Attaining Manufacturing Excellence

- Just-in-Time
- Total Quality
- Total People Involvement

The IRWIN *Professional Publishing* **/ APICS Series in Production Management**

Supported by the American Production and Inventory Control Society

Attaining Manufacturing Excellence

- Just-in-Time
- Total Quality
- Total People Involvement

Robert W. Hall

ISBN 0-87094-925-X

Library of Congress Catalog Card No. 86-71920

Printed in the United States of America

15 16 17 18 19 QK 0 9 8 7 6

PREFACE

A momentous change in the management of manufacturing has been underway in several parts of the world for more than a decade. How momentous it is and how far it will carry is for future historians to decide, but clearly the new approach is a departure from the practices of the mid-Twentieth Century.

This book is intended to guide middle and upper level managers of manufacturing companies on the nature of this change. *Selection of material was based on what it is thought a top level management champion of this change needs to know.* Consequently, the book was written as a readable text, assuming that the reader has a smattering of background in manufacturing.

To meet this objective, the book was written to show how the many facets of this management change fit together. Numerous topics are covered, but none in exhaustive depth or detail. An experienced engineer will think some of the discussion a little light, but a non-technical manager will not find the same material a quick read. Topics are covered in a sequence that by experience seems logical for gaining an overall understanding.

For a manager, the mental challenge is to create a vision of how to fit all the various pieces of this thinking into a blueprint for a particular company. In the interest of providing an integrative view, this book omits illustrations of setup time reduction, methods for performing a process capability study, methods for containerization studies, and other detail which might be of great interest to some readers. The first challenge is to gain an overall grasp sufficient to lead an organization through a broad scope of manufacturing change.

Thanks are due to many people who contributed in some way to this book. First, the writing was initiated as one of a series of books on manufacturing by the American Production and Inventory

Control Society. Their officers and staff have the foresight to promote changes in manufacturing that must be made.

Second, none of the concepts in the book are original, and everyone has a number of teachers. Some of the most effective teachers have been manufacturing managers in the U.S., Japan and elsewhere who have been kind enough to show me their factories and share their frustrations. Fragments of their stories are reported throughout the book.

A very effective teacher has been Professor Jinichiro Nakane of Waseda University, Tokyo. He and colleagues at the System Science Institute have been observers and students of excellent manufacturing practices in many parts of the world for a long time, and they have been willing to impart many of their observations.

Several reviewers provided very helpful feedback on all or part of the first draft of the book. The final result is much stronger as a result of their contributions. They are: Ken Stork, Motorola; Dick Newman, Indiana University Northwest; Ken McGuire, Consultant; John Burnham, Tennessee Technological University; and Jack Warne, retired CEO of Omark Industries.

Finally, my secretary, Jan Cashion, handled much of the word processing and detailed fussing necessary to create a manuscript in a home office also populated by two young children who provided no end of entertaining diversions from the work.

Robert W. Hall

CONTENTS

CHAPTER ONE

The Empty Strategy

American manufacturing dominated the world in 1946, if only because it was the only industrial base left substantially standing after World War II. In athletics too, Americans dominated the world. At the time few other countries could afford extensive athletic training.

That was 40 years ago. Other countries began rebuilding their industrial capability, and by 1970 it was unmistakable that players other than Americans were beginning to do well in the manufacturing arena. The 1970s were filled with dire warnings of comparatively low productivity growth in American manufacturing, and by 1980 the United States was facing tough international manufacturing competition, especially from the Far East.

Few would argue that U.S. manufacturing today is worse in absolute capability than it was in 1946. The key to the situation is that the rest of the world has advanced a long way since then. Both in manufacturing and in athletics, it is a different ball game.

The causes of the malaise in American manufacturing have been much discussed, often in terms of explaining why productivity growth has been slower than elsewhere. Numerous theories have been advanced, and there is probably some merit in many of them:

- The nations that rebuilt manufacturing after World War II have newer plant and equipment.
- American managers failed to reinvest in new plant and equipment, possibly because they were too concerned with short-term profit. To this accusation managers would frequently

reply that it makes little sense to invest heavily in flat or declining markets.
- The defense industry robs the country of its better technical talent. A money-is-no-object approach destroys incentive to improve productivity.
- Americans want high wage rates. This is frequently connected to union demands for high wages and restrictive job practices to spread employment among more people.
- Foreign countries use unfair trade practices in market access, sharing of technology, or in other ways. A counterargument is that manufacturers not feeling the press of foreign competition see no reason to make improvements and simply raise prices when they feel the need.
- It is hard to persuade talented young people to become technicians in manufacturing. They all want to get MBAs and not dirty their hands.
- "People do not want to work any more."

Though there have been some studies of these explanations, frequently such studies degenerate into political diatribes on whatever the speaker is against—real or imagined. (A Marxist will argue that the natural decadence of the capitalist system robs people of the will to work, which, after replacing some of the dogmatic expressions, sounds strangely like the "people don't want to work" theory advanced by managers who are clearly the opposite of Marxist.)

Manufacturing means different things to different people, so "saving" manufacturing also means different things to them. Until recently, few of the more general discussions of manufacturing difficulties focused on the fundamentals of manufacturing practice. That is a much more extensive discussion, and it takes longer to work through.

Terminology and perception differences plague any broad discussion of manufacturing. The words *manufacturing, production,* and *industry* are all used to mean almost the same thing: "operations creating wares having exchange value" in dictionary language. Here let us make a somewhat arbitrary distinction.

Manufacturing is *all* the activity of an operating company that engages in production. *Production* is the actual conversion of material to product. In a full manufacturing company, production is a subset of total activity—sometimes a subset employing only a minority of the firm's employees. A way to classify the difference is shown in Table 1–1.

The list of manufacturing activities in the table is short and simple, though perhaps not complete, and so is the list of production activities from a process view. The "organizational view" list is clear enough, but the production activities from the methodological view may seem strange. It is a bare-bones summary of a familiar subject from a different viewpoint—that of the classical industrial engineer.

Each list is really a summary of a viewpoint about manufacturing. And some viewpoints are confined to just a word on a list. For instance, a lifelong salesman may regard manufacturing mostly as a necessary support organization for a good line to sell. Likewise, inside the production function the various specialties may have little unanimity in point of view either, so they are not united on what to do to save manufacturing.

To most people saving American manufacturing means retaining some, but not all, production activities inside U.S. borders. Move fabrication offshore, and just assemble and test in the United States. Another idea is to just add one final production feature in the United States, as Toyota adds beds to imported pickup trucks after they arrive. A company can say that it retains domestic production even when that activity has dwindled to a shadow of its former self.

Some companies do not even have token production in the United States. They retain marketing and design in this country while relegating the production to other nations. *Business Week* recently dubbed this kind of company a "hollow corporation."

To a stockholder, saving a company means saving an investment, and if that comes at the expense of the production activity, so be it. To communities, it means saving a tax base and support for civic development; and for suppliers, it means saving at least a portion of their own business. To an employee, it means saving one's job. Unfortunately it may not mean saving the job as they have always known it.

From among the many perspectives and arguments over the past several years, the business press has begun to present articles on saving the production function by changing it. This story has come out piecemeal to most executives. Some of the changes seem pleasant; some do not. The changes are not an easy prescription

TABLE 1–1 Perspectives on Manufacturing

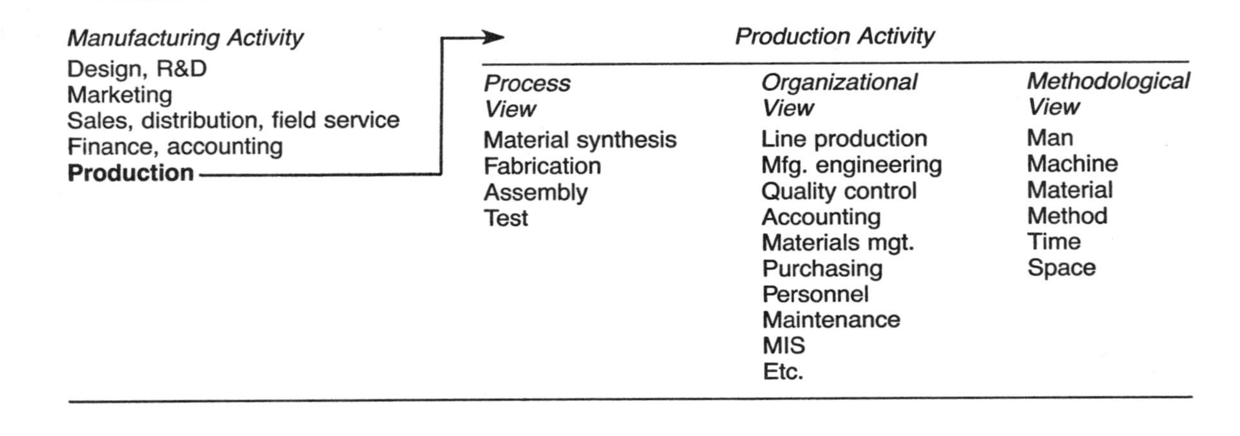

Manufacturing Activity	*Production Activity*		
	Process View	*Organizational View*	*Methodological View*
Design, R&D	Material synthesis	Line production	Man
Marketing	Fabrication	Mfg. engineering	Machine
Sales, distribution, field service	Assembly	Quality control	Material
Finance, accounting	Test	Accounting	Method
Production →		Materials mgt.	Time
		Purchasing	Space
		Personnel	
		Maintenance	
		MIS	
		Etc.	

taken as a whole, and they involve the entire thinking of manufac turing management from top to bottom—no mere technology change. Before taking this medicine, most managements must first resolve whether their production activity is worth saving.

WHY SAVE PRODUCTION?

A simple answer to this question is that it is necessary to develop knowledge and skill in production to be competitive in a full range of manufacturing activities, including marketing and technology. Any advanced economy needs to be strong in these areas. There are other reasons to save production:

- Continued employment.
- Support of national defense.
- Preservation of a healthy trade balance.
    - Warding off imports (defensive).
    - Increasing exports (offensive).
- Preservation of skills.

Saving jobs in the short run is a political motivation. An employed voter is more likely a happy voter. Whether considerable improvement of production will have this short-run result is open to question. In a highly competitive manufacturing world with flat markets, the necessary improvement of manufacturing requires considerable productivity improvement—fewer people. If productivity does not increase, companies fail or production moves offshore—and the jobs are lost anyway. One can thus conclude that manufacturing employment is more likely to decrease than increase in any scenario.

That conclusion should not suggest that manufacturing and its production activities are any less important. Agriculture has seen a steady increase in productivity and decrease in the numbers of people working in it for over 50 years. Eating is still important to everyone, whether they happen to be personally involved in food production or not. The same is true of manufacturing.

National defense depends on production. High-tech or low-tech, weapons must be produced. Nations with little production capability must use scarce resources to buy them from others, and what they buy is seldom the latest model. A nation that is able to design weapons, but not produce them with quality and quantity, also has a major weakness. Besides producing weaponry, a nation would like to be capable of sustaining its economy in times of crisis.

During World War II, production capability was obviously a very big factor. How much this is continued in an era of nuclear threat and terrorist action is subject to debate, but there is less question that political weight in the world is substantially affected by manufacturing capability.

In the 1980s preserving the level of manufacturing self-sufficiency that the United States achieved during World War II seems to be an unreasonable goal. With manufacturing capability much stronger worldwide, the United States cannot match every other region in every product line. However, production of a large stable of products competitive on the world market is necessary to maintain an *equitable trade balance* with other advanced nations. This begins with an ability to match imported products in essential areas.

Warding off imports demands attention to production fundamentals. As soon as counterfeiting manufacturers become able to produce product with quality indistinguishable from and sometimes better than the original, they are catching on. Ultimately they are able to develop and produce superior designs for the market.

Nations and regions use the development of production know-how as a stepping stone to technical and economic progress. Japan, Korea, and Taiwan are well-known examples, but many other nations also seek entry to advanced status by this route. Among them are such diverse countries as Malaysia, Finland and Brazil. It is obvious that each of these nations should be a good full-scope manufacturer of something to achieve the trade balance it seeks. These developing nations have no illusions of economic self-sufficiency at a high level.

Most nations following such a strategy first seek production of simple items and then more advanced ones, such as integrated circuits. If developing-nation manufacturers apply themselves and acquire the know-how, after a time they may advance to the technical big leagues, and from there to proficiency as a full manufacturing company with strong marketing power.

This is not a new strategy. About two centuries ago when the United States was a young nation, American financiers offered a bounty for the English Arkwright cotton mill technology. Young Samuel Slater, who had learned the mills in England, emigrated to this country and reconstructed Arkwright's mechanisms from memory, thus breaking an English monopoly and contributing to the development of the premier American manufacturing area of the time. The strategy was successful then and it still is. The story is much more pleasing if you happen to be American than if you are English.

By not paying attention to production methodologies a manufacturer invites raids by competitors. Some of these competitors will be foreign because they see such a strategy as essential to their balance of payments. But there is little reason why an aggressive domestic competitor could not use the same approach. If competitors strive to advance by being world-class in production, over the long run the attacked company must be close enough to world-class to keep pace with them.

When a company has become proficient both in quality and cost of production, it can aggressively go for export markets. In doing this, there are many important considerations besides production capability. But if a company is weak in production, its other strengths are sapped trying to cover for it. Thus it is difficult to gain much from exporting. A company that wishes to export a product will basically have to make most of it in the location from which the company wishes to export.

America itself is such a big market that many of its domestic manufacturers have not seriously considered export from the United States. In recent years, many firms have thought it so likely that other countries would erect protectionist barriers that these firms instead followed a financial and technical export strategy. They built plants in the market country or region so that much of their export/import was financial and technical. A decade ago the American auto companies were good examples of this strategy.

Now the large manufacturers of many countries have become multinational, and this leads to a need to compete with the best in several areas of the world. An American company building a plant in Mexico to serve either the Mexican or American market could easily find itself down the street from a Japanese company with the same idea in mind. The situation is a little blurred, but the basic importance of production capability is unchanged. A company manufacturing on a global scope must go head-to-head with the best.

The attention paid to production activities must be more than casual, more than just turning production over to technicians. Today's standards of production excellence are quite different than

40 years ago, or even 10 years ago. A gradual modification of traditional production practices may not be enough, just as a talented but untrained athlete may not last long in tough competition.

Manufacturing excellence is achieved instead when the skills in all functions of the company are integrated with skills in production. This alone is another reason for saving production. Preserving the production core of manufacturing is essential to saving the full scope of manufacturing activity, including marketing expertise.

Production capability is related to skill in design and development. For instance, the design of an integrated circuit is intertwined with the process for making it. It is one thing to conceive a technical idea for a product. It is another to determine how to design it for high-quality, low-cost production and for durable use. That is know-how. Without know-how, the entire manufacturing effort is weakened.

For much of its history, the United States acquired production skills from the rest of the world through immigration. And by its nature, the country allowed many of its creative citizens to develop further. Until recently, seldom have Americans said they traveled elsewhere in search of production technology (although the facts may be different). By contrast, Japan has had few immigrants, and the Japanese felt the need to search the world for the best they could find. Once they began to roll some of the best practices together with a few of their own contributions, the rest of the world wanted to know what they found.

As much as anything, the best Japanese companies (not the majority) sought fundamentals in production. No one in the world has a lock on those fundamentals, and everyone has difficulty mastering them. Fundamentals of production (interactions of man, machine, method, and so on) change little, though the technology by which they are applied may change. Technical advances are important to production, but technological strategies without careful attention to fundamentals is empty, just as the best football gear on an undeveloped player is worthless; it might as well be an empty suit.

Much of this book examines matters that should be fundamental to excellent production, and likewise to the management of manufacturing in its full sense. However, that does not mean that all fundamentals come naturally, or even that they are easy to understand. The practice of fundamentals is definitely not elementary. A good golf swing is fundamental also and superficially easy to understand, but for most of us it is neither natural nor easy to learn. In the same way, a company interested in top-level manufacturing competition must fully master fundamentals of production or risk having an empty strategy in competition.

To avoid having this empty strategy, a manufacturing company must strive for high-quality, low-cost, timely, and flexible production. That cannot be achieved by thinking of the production function as something to be managed in various specialized categories by technicians.

Why save production? After all, production is a part of a business, the primary goal of a business is to make a profit, and if that cannot be done, the investment should go elsewhere. Right?

Wrong. Survival is more than just an investment decision. Profit may be the goal of the providers of capital, but survival of production is of interest to more parties than the investors: employees, suppliers, governments, and communities. Survival by outsourcing is not seen as survival by these other constituencies of the company. In addition, production remains an essential element in the application of technology to useful ends—a source of economic strength. When a business makes a decision based only on the existing level of production expertise, it is giving up without much effort.

FULL STRATEGY AND SURVIVAL

Changes in production fundamentals change a corporate way of life. The degree of culture shock depends upon the status of the culture when change begins, but the reverberations shake the total outlook of management and its thinking about manufacturing strategy.

Businesses must base their strategy on what they know and play their strategic hunches on what they think they know. The hazard is in overlooking something important. Production capability should not be a blind spot.

Production capability is not the centerpiece of strategic thinking in Figure 1–1, and it should not be. Every business strategy begins with market, environment, and technical possibilities, but all the other elements are interrelated, and any strategy emerging from this milieu is useless without the capability to execute it. Leaders

FIGURE 1–1 Factors in Manufacturing Strategy—One View

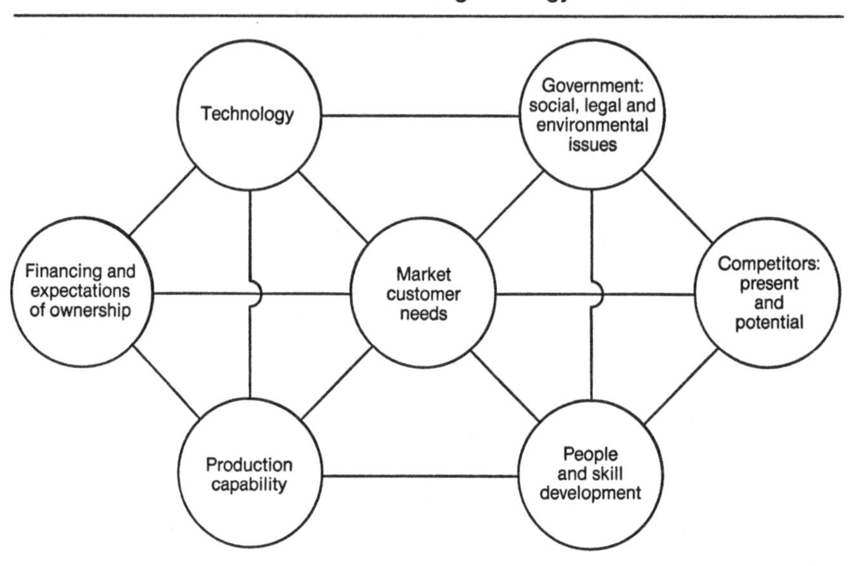

taking companies in new directions need broad exposure to many areas. One of these is production capability—what it takes to obtain it and what can be done with it.

Many otherwise broad-gauge leaders have had little exposure to production thinking, or what they have been exposed to was inadequate. Many managers with careers in production have not probed much beyond "how we have always done it" either, and those deeply implanted in production may not be comfortable with factors external to the company. Manufacturing companies seem to be encountering situations similar to that suggested by a financial analyst recently quoted in *USA Today.* Speculating on the resignation of a number two executive, he said, "They've decided they need someone more adventurous, more creative. They like ———, but he's an operations man."

The production function is too important to the execution of a full game plan to be considered as either dead weight or as something that can be tugged, hauled, and swapped about. To understand this, broad thinking business leaders must for a time become more immersed in production than many of them would like. They must analyze production for a while—study what production must do, how it must do it, and therefore what must be the necessary implications for people inside a company. Then they must go back and begin to think strategically. New production capabilities increase the possibilities externally, but at the price of operating with great skill and discipline internally.

Production capability is far from the only requirement for strong performance in manufacturing, but it does play a major role, and not only in low-tech businesses. For instance, IBM must have excellent production to compete with the horde of manufacturers of IBM-compatible equipment in the personal computer market.

As soon as IBM entered the personal computer market, it began to dominate it because of the firm's reputation in service and software. IBM's strategy in the computer business has always rested on selling service and solutions. In selling large mainframe machines, IBM did not have to emphasize rock-bottom production costs as long as the company could change hardware and software fast enough to give the IBM-compatible manufacturers a running target. These competitors were hard-pressed to match the service IBM lavished on large corporate accounts.

Entering the personal computer market jolted IBM's traditional make-to-order approach. The PC is made to stock and sold through distributors to individuals and small businesses. Direct contact with IBM's marketing and service representatives is broken, and cost becomes a bigger factor. These types of customers may like the broad range of applications software from IBM, but service representatives cannot hold the hands of all of them, so changing operating systems with regularity will not hold their friendship.

The compatible manufacturers were waiting for IBM to establish an industry standard. Then they needed only duplicate the capability of the hardware at low cost to provide a cut-rate machine that could use software written for IBM. Some have done an excellent job and field machines that some users actually prefer to IBM.

What kind of production capabilities do the compatible manufacturers have? Some are not so good. They just assemble components made elsewhere—where the real production capability resides. Others are very good. They need high-quality, low-cost production tied to the rest of the organization, with enough flexibility to quickly follow IBM should system architecture change.

As for IBM, it seems determined to remain the market leader and especially anxious to provide technical and service advances to the type of corporate accounts that have always been IBM's forte. What kind of production system does IBM need? Once again, it must be high-quality, low-cost, and flexible so that when the firm does decide to make changes in its machine, they can be quickly and flawlessly executed. To accomplish these objectives, the production capability must be well-connected to the rest of the organization. Should IBM by some flight of fancy ever reach the nirvana of selling software solutions and tossing in the hardware for free, it would still need that kind of capability.

Consider another example, that of manufacturers of diesel engines who supply manufacturers of trucks. Dependability, fuel economy, service, and cost remain important to customers, and several other factors could become important to the strategic situation of these firms over the next five years. Among them:

Current engine models have many complex options to fit various manufacturers' trucks and the applications of many truck users. Foreign competitors appear ready to attack them with high-quality, low-cost engines, but "plain vanilla" at first, forcing customers to rethink whether they need all the options they may have had in mind.

Low-cost manufacturers are making clones of their spare parts. Some are of good quality and some are not, but the diesel manufacturers feel the need to defend their service network and the aftermarket for spare parts.

Engine designs might change. Diesel manufacturers everywhere are known to be experimenting with ceramic parts which are lighter in weight or withstand more heat. Both of these properties contribute to fuel economy.

Computer-controlled fuel injection seems a certainty. From other sources, computer-controlled transmissions are on the way. What is the possibility of a computer-managed truck? How would such a development change the positions of numerous parties now prominent in the medium and heavy truck market?

Emissions levels remain a problem. If this concern is taken seriously enough, and if there is another oil crunch, truckers might want to switch to engines powered by natural gas. The United States has a large supply of that fuel.

There are still more uncertain factors in their strategic equations, but what kind of production capabilities do diesel makers need? Again, they require high-quality, low-cost, and flexible systems. They must be ready to acquire know-how rapidly in response to technical changes, and avoid making big investment commitments that could restrict flexibility in the wrong way at the wrong time. They cannot beat the competition by outspending them. They must match the competition in technology, marketing, production and especially people skill. To do this, they need improvements well beyond the trivial: major improvements in detailed quality leaving the factory, better durability and performance in the field, and cost reductions in the 20 to 40 percent range for current engines.

Whether the American manufacturers will be successful in this challenge has yet to be seen. If they are successful, they will be capable of working with a full strategy against competitors of a similar mindset. If not, production weakness can easily empty any strategy they attempt to follow.

To survive, a manufacturer needs some form of competitive advantage, if only a good location. (Protectionist legislation attempts to restore a location advantage to a manufacturer who has lost it.) Just as important, it cannot fritter away its competitive advantage by any major competitive weaknesses. The capability of its production must at least be a neutral factor.

To prosper, a manufacturer needs more than one advantage. A marketing advantage, such as flexible response to customers, cannot be sustained without production capability developed for short lead times. There are no perfect strategies, only effective ones, and effective ones are usually well-executed.

THE RISKS AND THE REWARDS

When undertaking a change in production fundamentals, most managements greatly underestimate all the facets involved. They underestimate the magnitude of the improvements to be gained, the quality improvements, the cost reductions, and the gains in flexibility. They underestimate the difficulty of changing management thinking. Ingrained habits do not break easily, and for all the simplicity of many changes, the emotional problems are immense. Most of all, firms implementing change underestimate the scope of the changes. Every part of a company is affected—marketing, finance, strategy—everything down to how the people of a company think of their place in the world.

As for the financial benefits of manufacturing revolutions few companies present a full financial disclosure of their efforts in just-in-time production, total quality control, employee involvement, and related matters. Most disclosures are succinct summaries typified by the following rough estimates from Hewlett-Packard's Vancouver (Washington) Division, one of the first American operations to venture into a combination of practices that might be called manufacturing excellence. Results cover roughly a four-year period:

Shipments	Up 20%
Work-in-process inventory	Down 82%
Space	Down 40%
Scrap and rework	Down 30%
Labor standard	Down 33%

A pioneer effort, this division probably could have done better had it been fully aware from the beginning what it was getting into. But the Hewlett-Packard plant was not. Management picked up ideas by bits and pieces before comprehending anything close to the full scope of changes, and despite the achievements, it still has many improvements to make. All this has happened while Hewlett-Packard's product line has turned over into a new generation. However, many manufacturing people will look at figures like this and say they are unbelievable.

How to go about implementing such change is the subject of much of the rest of this book. Many companies are able to achieve a small impact from a small change. Some can make great improvements through technological changes, although often at great expense. But top results come from an effort that is recognized as a complete change by an entire company learning to look at itself in a different way.

Manufacturing excellence is more than proposing a technical solution by engineering reasoning, then evaluating it financially by investment reasoning. It is more than proposing different behavioral programs for employees, but only hoping that these will somehow translate into improved performance by machines on material. One must put it all together—engineering, business, and people.

The hardest part is emotional—changing ourselves. Most of us (including the author) go through many stages on the road to understanding the composition of manufacturing excellence. These stages are different for different people, but the first reaction usually is that few of the stories are true. Then the thought is that nothing about the changes is new or different, and there is some truth to this criticism. Then the whole approach looks too simple. Finally, it all begins to look simple in concept but requires tremendous dedication from people, and the actual doing is not always so simple.

Understanding manufacturing excellence is more than an intellectual exercise, though the emotion level stops somewhere short of revivalist fervor. Major improvements in production capability require substantial intellectual and emotional commitment on the inside of a total corporate organization, and some degree of support by outsiders who influence it.

Incidentially, though Japanese may have some cultural affinity to perfect manufacturing, the whole matter is not a question of national culture. The author knows two Americans who have been dispatched to Japanese affiliates to assist them in starting "Japanese manufacturing."

Conventional business thinking cannot cope with the total breadth of the concept of manufacturing excellence. Recently I assigned to a group of MBA students a well-known case, North American Rockwell Draper. It is the story of a foundry in decline serving the textile industry, which was also in decline. The foundry had overcapacity, lack of flexibility in its newest equipment, pollution problems, and potential liabilities from strained backs.

Almost all the MBA's were working students, and almost all opted for the "conventional out"—that is, for investing the minimum and riding the business down. A sample comment from the case writing: "You cannot invest enough to buy your way out, so wring whatever you can from an old operation until you have to ask for concessions from the work force. (The work force had little to give back.)"

These MBA students had no foundry experience, and no foundation from which to propose working in any fashion toward a fundamentally different approach to production capability. They were stuck doing present value analyses of the alternatives described in the case, and their recommendations were based on investment thinking. Nothing prepared them for leadership thinking in such a situation, which was not a fact in which their instructor could take much pride.

Leadership thinking is very necessary, along with development with patient investment. A company can make a 10 to 30 percent cut in inventory and think it has attained a great accomplishment in short order. Space reductions come quickly if equipment is mobile and no one has previously given thought to compact layouts. On the other hand, the benefits of improved quality and higher productivity come slowly, for these depend not on any fast-response technique, but on the patient development of people. Flexibility in manufacturing is not something easily measured, and sometimes its presence or absence is not a matter that management is consciously aware of.

Managers are wise to financially summarize the prospects of a course of action to check whether it holds any promise of profitability, but no one can make an informed judgement about the prospects for manufacturing excellence by compiling and evaluating financial projections. The issues are people, products, and processes—leadership and development. Making such an early evaluation is much like projecting whether a promising high school sophomore athlete can be developed into a major leaguer.

This is not a recommendation to take leave of all business prudence and stake everything in a company on a wild leap of faith. That is just another search for a magic solution. However, it is true that a number of companies have not found the courage to make a strong break with past manufacturing tradition until their condition was so desperate that they had to leap without being completely sure of their landing spot. That is as true of as many Japanese companies as Western ones.

THE SURVIVAL PERSPECTIVE

A large manufacturing company in imminent danger of collapse, such as Chrysler in the early 1980s, is a matter of interest to more than just its stockholders. The Chrysler near-debacle was well-chronicled, and the firm's rescue was aided by all its constituencies:

1. *Customers* loyal to Chrysler placed advanced orders for Aries and Reliant cars before they had even seen them. Lee Iacocca's televised appeals to help an American institution swayed enough people to advance sales that even the banker of last resort, the U.S. Senate, was inspired to intervene.
2. *Creditors* stretched themselves through a long, complex set of arrangements and guarantees. Hundreds of jittery lenders

participated in loan packages headed by "keystone" bankers who took the risk that Chrysler would survive.
3. *Employees* paid the price, some with their jobs and all with pay deferments.
4. *Suppliers* stretched themselves by extending terms and by improving their quality of service during the crisis.
5. *Communities* rallied to prevent the loss of Chrysler plants and Chrysler jobs.

Chrysler did come back from the brink, repaying the federal loan ahead of schedule. One of the reasons was that it found itself able to operate with about $1 billion less than expected in inventory because of a set of practices that the auto industry began to call just-in-time production. The excess material came from every part of the company, beginning with a simple inventory work-down. For instance, a Chrysler foundry was simply ordered to pare the on-hand quantity of castings from 15 days to 5. They never missed it.

For many inside Chrysler, this epic is not over. Some, however, think everything is nearly back to normal. One Chrysler employee was heard to say, "These guys made a billion dollars last year, and some of them think we have really accomplished something. That scares me." Call that the survival perspective.

The perspective is necessary to understand the new approach to manufacturing. From the Chrysler story and several others, many in the manufacturing community understand this new approach in a superficial, nonintegrated way. To some, just-in-time is "closely timed shipments that permit reductions in inventory." (That is only a small part of it.) Total Quality Control is a "broad application of statistical process control." (There is more to it than that.) Quality of Work Life is an "effort to make employees feel they are participating"—an obviously hypocritical statement.

Each of these statements reveal that their speakers are thinking of management as a set of techniques. But such techniques are not mastered, or even correctly understood, without adopting the spirit of them—the survival perspective.

That perspective unifies goals. It provides an integrated view of manufacturing. Techniques do not work unless the various constituencies with interest in the company work together. They need not all go to the same church, or socialize together much. But they do have to work together.

During its financial crisis, Chrysler's plight attracted great attention. The story inspired sacrifice because it was a fight for operating survival—something everyone could identify with. But not everyone and everything survived with Chrysler. Products, processes, operating methods—and people—all changed. There is still much more to do, but a reduced sense of urgency about it.

Lee Iacocca's salesmanship continues to keep Chrysler products in public view, and the marketing campaigns are vital in sustaining the company. A step-by-step advance toward improved production capability is also vital to Chrysler's future, but it is hard to maintain a spirit of sacrifice for such a seemingly undramatic goal. This story receives uneven news coverage, and it goes much more slowly than many at Chrysler would like.

Elsewhere, other companies struggle with their own halting pursuit of manufacturing excellence. Aside from those attempting to save production, few realize the dedication necessary. Progress often seems to be two steps forward and one back.

The proper attitude for these manufacturers is a survival perspective, because excellence is not attained with a few swift steps. It is easy to accept the platitudes. Survival involves much more: concentrating on the customer, attending to detail, ad infinitum. In fact, one hazard is becoming bored with some of the detailed improvements necessary—for example, rigging a machine to be fail-safe in grinding a surface, developing a way to clamp a wire in exactly the same way time after time, or reducing the travel distance of a part by six inches. These are not accomplishments for which brass bands march. But many little advances by many people add up to a big result.

Manufacturing excellence results from a dedication to daily progress. Make something a little bit better every day, including every employee's skill. It takes a sense of direction to know what is better—specifically what represents better quality, less waste, and faster response to changes. Manufacturing excellence strives to improve activities that contribute to customer well-being in ways often unseen and frequently unappreciated.

Developing a sense of direction is vital. There can be many distractions. Computers seem a way to greatly improve production activities. Automation seems a way to eliminate many troubles. That may or may not be true, depending on how these tools are applied to the production situation. The inclination is to develop plants to take advantage of economies of scale, using equipment to run items fast, in large lot sizes, and at low cost. That theory works as long as competitors are not making the products in small lot sizes at speeds matching market demand and at still lower cost. In so doing, they may have better quality, too.

Everyone wants to compare American performance with Japanese. The observations of Professor J. Nakane and others who have seen leading manufacturing operations, both American and Japanese, help put this situation in perspective:

1. New American plants often have excellent technology that is at least equal to Japanese and usually better. That is, any technology gap still slightly favors Americans.
2. American computer systems and software are almost always superior to Japanese. They are larger, more complex, and more powerful, but this can be a weakness as well as strength if the systems mask wasteful practices that should not exist.
3. Japanese are almost always superior in their ability to improve existing plant and equipment: tooling improvement, defect elimination, layout improvement, and so forth.

The conclusion is that Americans have trouble putting the pieces together and making the most of what they have. For some firms that is necessary for survival, and for all it is necessary for excellence.

At first the hardest part of this challenge is believing that anything should be done. Then it becomes difficult to stay motivated long enough to accomplish anything. A few people are intrinsically motivated to continuously improve manufacturing, but most are motivated only by occasional nightmare images of rugged competitors swooping down on a treasured market. Motivation is stronger if the images are stronger. What are the production capabilities of competitors, and how does that relate to their ability to capture market share?

Large companies refer to this as benchmarking, or baselining, themselves and the competition. What are the best production capabilities of an industry in key areas, and who has them? Some suggestions of capabilities to benchmark are:

- Product quality.
- Production defect rates.

- Equipment setup times and resulting lot sizes. (This has to do with flexibility.)
- Lead times: For material.
  For customers.
  For product mix changes.
  For new products.
- Inventory levels.
- Productivity levels.
- Maintenance (unplanned downtime).
- Size of supplier networks, and supplier capabilities.

Gathering and interpreting such information is not always simple. Even if companies freely part with such data—and many do—the configuration of production processes to build competitive products is not always the same, so that matching such data between companies becomes difficult if differences are small. Most companies do not readily part with internal cost information, but even when they do, interpreting comparisons of costs is subject to the same problems.

This is not the point, however. Big differences between companies have importance regardless of interpretive subtleties. One can be sure that if one company has the competition beaten in several key categories, its ability to execute is superior, and very likely its costs are lower.

The next step is to compare these capabilities with the apparent strategies of each competitor. A company with superior production capability can still be broken if it is weak in marketing strategy, or in some other way. Production capability is only one category in an overall assessment of competitors, but it is a very good indicator of their potential if they are able to use it. If it is good, a competing firm can execute a strategy well and thus is not an empty threat.

All developing nations have people skilled in tooling, maintenance, layout, and other skills essential to attaining manufacturing excellence should they be motivated to try for it. It is a dangerous illusion to believe that these foreign firms, or well-motivated domestic competitors, can be overcome by business as usual—that is, by pumping out goods using established methods. A company seeing potentially tough competition must "go for it"—manufacturing excellence.

Manufacturing survival is war, combat on speaking terms perhaps, but war nonetheless. Every resource must be in readiness and every skill at full alert. True, some reasons for the difficulty American-made goods have with foreign competition are labor rate differentials, trade restrictions, national promotion of various industries, and so forth, but at some point manufacturing competition comes down to basics, the infantry tactics of industrial management. The company that cannot compete in this battle cannot forever be defended. The real defense is to think aggressively of what it takes to compete in a world market: world-class manufacturing, even if the company is too small to actually dent world market volumes. If a thriving company waits for excellent competitors to attack before becoming worried, it will sit in comfort until this event really happens.

Development of production capability is war on the "home front." The first step is recognition that something must be done. This must be followed by rededication to the art of manufacturing. Rethinking of techniques is necessary, but not enough. Concern by production managers is necessary, but not enough. Manufacturers with all their constituencies must have some degree of concern. Manufacturers would do well to ponder one of the precepts of Toyota Motor Company, "Win first. Profit later." Though a little obtuse, this slogan is very indicative of the attitude necessary to build superior production capability and thus avoid an empty manufacturing strategy.

CHAPTER TWO

The Philosophy of Value-Added Manufacturing

The "new approach" to manufacturing is a pragmatic philosophy distilled from worldwide experience in manufacturing. The major concepts are independent of technology, though they may be applied differently with technical advances. Taken independently, none of the concepts are new; all have antecedents dating to the early 20th century, if not before.

Those long in manufacturing experience who feel that they have heard it all before have a point—but one that may deceive them into missing the major point: the novelty of thinking is to combine the best and simplest practices one can find into an elegant whole for a given application. Anything that broad in scope is a philosophy.

Begin with the objectives:

- Eliminate waste.
- Reduce lead times for:
  - Customers.
  - Materials.
  - Tooling and engineering changes.
  - New product introduction.
- Increase quality.
- Reduce costs.
- Develop people: Increase skill, morale, and productivity.
- Improve continuously.

The goals are broad and ambitious. Attaining them demands basic, fundamental improvement in all aspects of manufacturing. The mind should not focus narrowly on a particular technology or technique but should open widely. Many problems stem from nonunderstanding or nonacceptance of *all* goals, as in believing that the objective of something called JIT is only to cut inventory. Trivial goal; trivial achievement.

None of the names often used for this philosophy suggest its total power and scope. Semantics cannot suggest what the mind has never grasped. The significance of any name only materializes after exposure to the philosophy itself. Many names infer a set of practices narrower than their users intend:

1. *Just-in-time:* Suggests a program limited to scheduling, inventory, and logistics. The goal seems limited to cutting inventory.
2. *Stockless production:* Bland and still suggests concentration on inventory.
3. *Total quality control:* Sounds like super-modified statistical process control.
4. *Zero defects:* Sounds too idealistic to believe.
5. *Zero inventories:* Same problem as zero defects.

Four names designate the philosophy throughout this book. None are universally used, and all relate to some aspect of the philosophy.

1. *Manufacturing excellence:* One way to suggest improvement in its broadest context. "World-class manufacturing" or other superlatives may suggest a combination of the very best.
2. *Value-added manufacturing:* Do nothing that does not add value to the product or to the customer. One of the precepts.
3. *Continuous improvement manufacturing:* Every aspect of manufacturing is dedicated to making it better in ways great and small.
4. *JIT/TQ:* This abbreviation suggests a combination of just-in-time, total quality control and whatever else a company is doing that is a significant departure from its previous approach to manufacturing. An abbreviation such as this is probably the most popular name. (After the philosophy is firmly in place, it no longer needs any special name.)

Most companies honestly following the philosophy describe their approach in three overlaying categories of work, as suggested by Figure 2–1. One cannot be successful with just-in-time manufacturing or total quality without total people involvement, no matter how automated the process is. A few phrases in the boxes suggest major concepts in each category, but some concepts should be listed in more than one box.

Most explanations of JIT, TQC, and related topics begin with lists of points, somewhat as in Figure 2–1. None of the lists are, or can be, complete. Each list is really intended to summarize a point of view. The subject cannot be mastered as a body of closed-system techniques, although understanding techniques helps to understand the philosophy. The best understanding comes from doing because manufacturing excellence is really a philosophy of doing. Just reading and talking about it is "Mickey Mouse."

The name *value-added manufacturing* is derived from the objective of eliminating waste. The definition of *waste* is: Anything that does not add value to the product or service, whether material, equipment, space, time, energy, systems, or human activity of any sort. Purging whatever does not add value implies careful attention to (1) establishing what *does* add value to the user of the product and (2) identifying whatever *does not* add value.

Careful observation of user wants and needs is everyone's job, but it is the prime business of market research and of those in customer contact. Their observations should detail value added to the customer in a form useful for engineering and production. This is important for deciding which product characteristics are critical for quality attention and what the specific requirements should be.

Shigeo Shingo encapsulates identifying and eliminating waste in his summary of the seven wastes (Table 2–1). The seven wastes should be savored and mulled over because they all interrelate, and they are easily hidden in the complexity of a large organization. For instance, Apple Computer developed a large holding area to monitor Macintosh computers during burn-in. After considerable work improving quality, the first-pass rate at functional test rose to 98 percent, so Apple abandoned the burn-in test because it added little value. (How many of the seven wastes are illustrated here?)

Shingo thinks of production in a physical sense. His background is in industrial engineering and shopwork, and his methodology of observation is simple. Look at the operations directly and ask why each is necessary. Were he more familiar with Western manufacturing, Shingo might have added an eighth waste: unnecessary measuring, recording, and managing in an effort to deal with unnecessary

FIGURE 2–1 Overview of Manufacturing Excellence: Value-Added Manufacturing

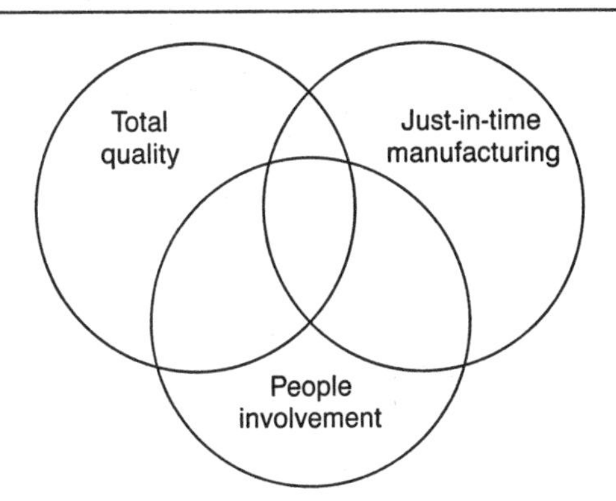

TQ = Total quality	PI = People involvement	JIT-M = JIT Manufacturing
Defined quality to customer	Survival perspective	Workplace organization
Total company effort	Total organization reform	Visibility
Targets for improvement	Responsibility at the source	Limited inventory
Quality process; quality product	More skill, less effort	Reduced setup times
Responsibility at the source	Flexible workers	Small lot sizes
Statistical process control	Broad perspective	Reduced lead times
Immediate feedback	Full work	Reduced space:
Cause-and-effect methods	People development	Group technology
Reduced variance in process	Problem-solving atmosphere	Standard routings
Failsafe operations	Performance measurement	Cell layouts
Standardization	Continuous improvement	Producible designs
		Stable, repeating schedule
		Preventive maintenance
		Cycle time analysis

complexity. Wastes observed in physical operations often provide clues to waste far removed from the direct activity.

Most readers at first have little patience to reflect on the seven wastes. Two aspects of them cloud understanding: (1) They derive from a physical vision of the business world, and (2) they are holistic. Try to visualize physical entities; verbal symbolism inadequately describes the "left hemisphere brain activity" from which some of the concepts originate. Then think in physical terms of the complete process of providing the customer with a product that fits his purpose. Value added manufacturing emphasizes the why and how of serving the customer, derived from direct observation of operations.

The seven wastes did not originate with Shingo. Taiichi Ohno, an early advocate and developer of the philosophy at Toyota, pre-

TABLE 2–1 The Seven Wastes

1. *Waste of overproduction.* Eliminate by reducing setup times, synchronizing quantities and timing between processes, compacting layout, visibility, and so forth. Make only what is needed now.
2. *Waste of waiting.* Eliminate through synchronizing work flow as much as possible, and balance uneven loads by flexible workers and equipment.
3. *Waste of transportation.* Establish layouts and locations to make transport and handling unnecessary if possible. Then rationalize transport and material handling that cannot be eliminated.
4. *Waste of processing itself.* First question why this part or product should be made at all, then why each process is necessary. Extend thinking beyond economy of scale or speed.
5. *Waste of stocks.* Reduce by shortening setup times and reducing lead times, by synchronizing work flows and improving work skills, and even by smoothing fluctuations in demand for the product. Reducing all the other wastes reduces the waste of stocks.
6. *Waste of motion.* Study motion for economy and consistency. Economy improves productivity, and consistency improves quality. First improve the motions, then mechanize or automate. Otherwise there is danger of automating waste.
7. *Waste of making defective products.* Develop the production process to prevent defects from being made so as to eliminate inspection. At each process, accept no defects and make no defects. Make processes failsafe to do this. From a quality process comes a quality product—automatically.

SOURCE: This is a brief summary and commentary on the seven wastes described by Shigeo Shingo in *Study of Toyota Production System,* (Tokyo: Japan Management Association, 1981), chap. 5, p. 287.

sented, and elaborates on, the seven wastes in his 1978 book.[1] Ohno also is a "left hemisphere type" who said that even after reaching top management he preferred to receive direct, live information by observing plants and other operations, not by reports or meetings in his office.

Ohno and other Toyota managers in turn do not claim that their thinking about waste is completely original. Ohno's book has a chapter devoted to Henry Ford I, whose ideas on waste are not thought to be original either. They were derived from his own labor theory of value:

> My theory of waste goes back of the thing itself into the labour of producing it. We want to get full value out of labour so that we may

[1]Taiichi Ohno, *Toyota Production System—Beyond Management of Large-Scale Production* (Tokyo: Diamond Publishing, 1978). (In Japanese)

> be able to pay it full value. It is use—not conservation—that interests us. We want to use material to the utmost in order that the time of men may not be lost. Material costs nothing. It is of no account until it comes into the hands of management.
>
> Saving material because it is material, and saving material because it represents labour, might seem to amount to the same thing. But the approach makes a deal of difference. We will use material more carefully if we think of it as labour. For instance, we will not so lightly waste material simply because we can reclaim it—for salvage involves labour. The ideal is to have nothing to salvage.[2]

From this view evolved the more encompassing concept of waste as *anything* unnecessary. In highly automated plants, the labor theory of value becomes somewhat diluted, but wasting robot-hours has no more point than wasting worker-hours. The passage also brings out Ford's motivation to pay labor its full value and strongly hints that his thinking was based on an ever-increasing skill in doing work. The art was to see through the muddle to the how of it, and that he did by maintaining a strong physical vision of what his company and the world should be like. (The same book suggests that the time for Model T engine parts to flow through the Fordson plant was roughly the same as that achieved by Toyota today.)

Like Ohno—and hundreds of others who have contributed to this thinking with their experience—Henry Ford I was a left hemisphere thinker with a well-practiced disdain for those with overly abstract views of manufacturing—notably accountants and engineers. Also like Ohno, he had both an eye for detail and a vision for the big picture, an ability difficult to cultivate in practice. Ford was intimately familiar with the machinery and materials used for the automobiles (and many other products) of his day, and his ambition was to develop the human institutions to bring about the technically possible.

Several "principles" of value-added manufacturing reappear throughout this book in many ways.

Take a Broad View of Operations. The intent is not to create a narrow, production-oriented organization but rather to create in

---

[2]Henry Ford I , with Samuel Crowther, *Today and Tomorrow* (New York: Doubleday, 1926), p. 91.

everyone a sense of serving the customer. To that end, no established practice should be considered sacred. A supervisor in the Kawasaki Motorcycle Division in Akashi, Japan, was questioned about how the work force responded to a bobble in production. "We build the world's best motorcycle," he began. All the detailed explanation fell under that heading.

Make Problems Visible to Everyone. Make waste visible so that many ideas to eliminate it completely can be conceived and tried. This approach runs counter to the instinct to hide trouble and appear to be doing better than is actually the case. An illustration is the purchasing manager who made the common error of thinking that the main objective of just-in-time production is to reduce inventory investment. He pressured a supplier into holding parts for delivery on short notice but without increasing prices. For a time he fooled his boss, his accountants, and himself, but not the workers—who could read old production dates on the tags coming in with the parts. The waste, or rather a major symbol of it, had been transferred in ownership and location, but it actually increased. Just as much production work and even more double-handling of the parts was required. It all came out when a major quality problem appeared. Customer and supplier had to face the truth in discussing responsibility for the loss and what to do about it. Supplier physical operations are part of the total picture; they just happen to be external to the customer plant.

Keep It Simple. Persist to the core of a problem. Ask yourself why five times, and you may get close to the truth. Better, if many people ask why five times, truth is even more likely to come out and be in a simple form. Much complexity comes from an incompletely resolved problem.

Well-known examples of simple solutions are many: Henry Ford's famous crates with holes drilled and countersunk so that the boards could be used for Model T running boards; the fellow who suggesting deflating the tires in order to free the truck wedged under an overpass; AT&T sending telephones to the service stores with different colored cases kitted separately from the telephone mechanism, enabling customer combinations to be assembled in the store, which improved customer service while simultaneously decreasing field inventory. (The practice was discontinued because assembly became difficult for clerks.)

Unfortunately, not every problem has a simple solution; but many of the great solutions are simple, born of thinking of the problem in many different ways. Simple solutions work better. When they are not found, perhaps we are still not at the root of the matter. Keeping it simple is not simple. It is an admonition to extreme thoroughness and clarity in thinking at all organizational levels.

Improve Operations before Spending on New Plant and Equipment. An electronics company used a $250,000 laser cutter for complex-shaped printed circuitboards. Volume increased until it was close to the cutter's capacity. The obvious solution was to cut multiple thicknesses of the same board material into the same shape with a single cut, but the laser heat fused the cut edges of the layers. Justification for a second cutter was being prepared when an operator discovered the "obvious" solution by trial and error. A layer of paper inserted between the boards prevented them from fusing.

However, this principle runs much deeper than just searching for alternatives before buying a new piece of equipment. Simple operations start with a simple product design. In the case of the complex-shaped circuitboards, an unanswered question was whether to use a simple-shaped board that could have been die-cut with less expensive equipment. (The design came from the customer in this case, so the question was never pursued but could have been.)

The principle still applies in developing a total operation. For instance, eyeglasses once were made in shops by optical craftsmen who also fitted them. Later, glass was ground and spectacles assembled to order in centralized factories, after which they were sent to the optometrists who fitted the prescribed glasses. For the past 20 years or more, the trend has been back to custom assembly of common prescription glasses at the point of examination, often with a wait of an hour or less—a form of just-in-time delivery. The "eyeglass factory" concept depended on the total concept of serving the needs of those who require glasses. And all this has been accompanied by controversy within the field of optometry, of course.

This principle applies to all manufacturing decisions about new and revised facilities. Try to gain as much insight as possible from detailed, firsthand experience in rethinking and revising operations as they should be before investing much capital.

Another example is a furniture company whose business grew until it felt an automatic warehouse to keep inventory under control was needed. When that warehouse filled, another was added, but for lack of real estate, it was located about a mile away. Material received there moved into the plant as needed. Now the company is busy with scheduling, quality problems, and material flow in the factory to eliminate need for not only the outside warehouse, but both of them.

The key is relentless and continuous attention to the operation's detail by *everyone* in the company. There is no magic, but rather a great development of many eyes for spotting waste and a dedication to eliminating it.

Flexibility. This term is used in several ways, all of which boil down to keeping the options open if the future does not work as planned.

1. Flexibility to survive variations in volume. Use ingenuity to keep the cost of plants, equipment, and operations low so that operations have a low break-even point. (Toyota is said to ideally maintain a plant so that its break-even point is no more than 30 percent of rated capacity The ideal also militates against free spending in capital projects without careful study of the needs and purposes of the operations.)

2. Flexibility to change model or product mix. This ability comes from flexibility of several types: flexible workers who can do many different jobs, short setup times on equipment and tooling, and a low inventory of a wide mix of materials.

3. Flexibility in use of equipment. First, attempt to adapt general-purpose equipment to a specific application. If special equipment must be built, try to keep it inexpensive. Where weight and necessary operating stability permit, try to maintain equipment so it can be moved with minimum expense. (This is the same principle as office buildings with easily moved interior partitions and equipment on casters.)

Focus of manufacturing has been advocated since Wickham Skinner introduced the idea over a decade ago.[3] Flexibility and

---

[3]Wickham, Skinner, "The Focused Factory," *Harvard Business Review,* May–June 1974, pp. 113–121.

focus seem contradictory, but they are not. Focus is the ability to concentrate on the necessary tasks at the necessary times. Flexibility is ability to quickly change focus. Flexibility in layout, worker skill, quick-change tooling, machine conversion ability, and the like actually improves focus. Excellent production capability demands both flexibility and focus.

Loss of concentration on necessary tasks fouls operations: orders are taken that do not fit the work methods, or workers have difficulty understanding what is or is not important. Worse is continuous uncertainty about what to do or how to do it—lack of focus.

A metal fabrication shop mixes fabrication of construction steel with fabrication of aluminum rails and frames that need cosmetic appeal when finished. Both are sometimes finished together in the same area because schedules collide and the shop is small. Steel moved or rolled over aluminum scuffs it. Flux and oxides from steel sometimes fly far enough to spatter the aluminum. Workers try to be careful, and job schedules are such that steel and aluminum are not worked together, but when jobs are delayed or expedited by customers, these rules do not hold.

Why do they do it? The shop needs money, so it is trying to build up a business in aluminum. If it gets enough business it will separate the work, but meanwhile employees are working overtime to combat waste and quality problems. The shop has flexible workers and equipment but not a layout flexible enough to focus.

Many problems are just this basic, even where technology is more to the high side. It is a matter of tough business decisions, not just technical ones; in the case of the fabrication shop, the layout problem is difficult in cramped space. The shop wants to attract enough business in either aluminum or steel to be in one or the other, or enough to split and be in both. A condition in which the bank demands a statement every month allows no formula for focus, only skill and hustle pursuing a decision to move out of the status quo.

ENGINE PLANT EXAMPLES

Engine plants illustrate all the "principles" discussed so far. An automated line can be built to automatically transfer blocks from machining station to machining station. If it is built to machine V-8 blocks, it cannot be modified to machine sixes or fours, so such an investment is a gamble that V-8s will remain popular. Modifying a transfer line specially built for in-line six-cylinder engines to machine in-line four cylinders also may not be a trivial undertaking.

If the equipment has only two speeds, on and off, the only way to adjust to different production volumes is with total "on" time of the equipment. If the machine is not modularly designed, a malfunction in one section requires tear down or adjustment in other sections. Inflexible special-purpose equipment is a thing of beauty when doing what was intended, but ugly if intentions go awry. Complex equipment has more ways to malfunction, and repair is more complex.

Suppose the company also automates transfer lines without cleaning up the quality of the operation itself. Castings must sometimes be removed from the lines because of cracks or porosity (often unseen until after machining). For one reason or another, the automation needs much attention. This picture, while considerably simplified, has unfortunately been painted indelibly at least one time in the production experience of almost every major engine manufacturer.

For many years, the Camelot of this manufacturing philosophy has been the Toyota Motor Company's Kamigo Engine Plants, a complex that produces engines and transmissions. In 1981, one of these produced 1,500 six-cylinder engines per day, using a total work force of 161 people on two shifts. Most of the people work in assembly; the rest tend and maintain equipment or do material handling. Production averages about nine engines per person per day, well above the productivity of any other known automated engine plant.

Going through the plant, one can believe the productivity figure, but to the uninitiated the scene does not fit the expected just-in-time plant or marvel of automation. The Kanban card system is only used to bring parts into the plant. There are (or were) no robots, little programmable equipment of any kind, and almost no new equipment. Much of the equipment is 50s and 60s vintage, general-purpose machines modified for a special purpose; thus, it would not take much to unmodify them, if need be. Special equipment appears "home-grown" and inexpensive.

Transfers of material between machines is automatic on simple powered conveyors and transfer slides. Their movement is regulated by workpieces hitting "flappers," which activate microswitches. The system allows machining at variable rates up to the maximum permissible by the machining processes themselves. Pull more finished parts out at the end, keep material fed, and the network of machines increases the frequency of the work cycles they perform. Stop pulling finished parts out, and the machines stop themselves. What more effectiveness could one want?

However, the tendency is to turn up the nose at this plant, perhaps as quaint, or reminiscent of a Rube Goldberg creation, though one suspects that Henry Ford I would have been proud of it. Early in the century he once described River Rouge machine shops as packed with a higher density of general-purpose machines than any other plant on earth.

One Kamigo manager was asked why the handcrafted "flapper" sensors are still in use. Why not photocells, or pressure sensors, or magnetic detectors, or whatever? "There is no point in them. The mechanisms now in use have worked since 1966, and such changes would add only to the expense without improving the result—a waste. We will change immediately if it will produce a better result."

One Kamigo plant is compared with two "average" plants in Table 2–2. No figures are available, but the Kamigo plant obviously has far lower investment. The differences are startling but often explained away on the basis of different conditions, etc. The excellence of a plant depends not on its technology but on how well the total mission is accomplished by the quality of the total operation. The Kamigo plant works because quality is good, machinery and tooling are maintained, workers are well trained, management has both short- and long-term flexibility with respect to the missions expected of the plant, and they spend very little to do it. Perhaps the same sign could be hung over the Kamigo plant that is attached to the front of a 1950 pickup truck driven by an elderly man in my neighborhood: "Paid for and written off. Stay the hell out of the way."

This example provokes skepticism because the Kamigo plants produce a repetitive product. (Their engines do have a large variety of final dress configurations.) Not many plants have such a mission, and as widely discussed, the nemesis of Henry Ford I was General Motors' ability to produce a greater variety of cars. Henry's miscalculation was that 95 percent of the market would be satisfied with basic transport painted black. More variety adds complexity,

TABLE 2–2 Comparing Engine Plant Capabilities, 1984

	Toyota Kamigo *	*Chrysler Trenton*	*Ford Dearborn*
Products	2.4 L4 2.0 L4	2.2 L4 Incl Turbo	1.6 L4 HO, Turbo, EFI
Plant size (square feet)	310,000	2.2 million	2.2 million
Hourly employment	180	2,250	1,360
Line rate (per day)	1,500	3,200	1,960
Labor-hours per engine	.96	5.6	5.55
Shifts	2	2	1 assembly 2 machining
Total average inventory	4–5 Hours	3–5 days	9.3 days
Robots	None	5	n/a

*Variations and fragments of this data are known to have circulated for about five years. A table with these figures appeared in complete form in "Quality Goes In Before the Part Comes Out," *Automotive Industries,* November 1984, p. 52. Some of the numbers on the Kamigo plant vary slightly from those the author earlier collected personally, but they are not substantially different. The American plants have more complex engines, not all the space is used just for engines, and the operations performed do not correspond exactly. The American shortfall in productivity is not as drastic as the numbers show, but it is still big. Several speakers have used versions of this data for shock effect in presentations.

but it does not obviate the need to eliminate waste. Variety without waste then becomes the goal.

How is it done? By concentrating on basics, which are not always so basic. Just as is true in athletics, mastery of fundamentals comes first. Even veteran professional athletes begin their training seasons by concentrating on fundamentals, and when games go poorly, they return to them for review. This subject is really about fundamentals—or what should be fundamentals, given that manufacturing is not an expensive sport but rather a serious business of survival.

TOTAL QUALITY CONTROL

Begin with the customer. Quality in the technical sense must contribute in some way to customer satisfaction. Customers need know only how to use a product properly, and they may or may not give much thought to whether they are satisfied with it. Few technically analyze their needs and habits, but a quality manufacturer must do that with excellence in order to design, build, deliver, sell, and service the product—translate customer needs into the technical requirements and organization detail necessary to perform.

Measure and track quality. Display the history of key quality measures prominently. Measures are sometimes difficult to interpret (for instance, user perception of color fade), but all too often measures that should be instinctive are not known. When visiting a plant, ask for data like the defect rate finishing final assembly, a measure that should be well publicized in some form. The common response is: "Oh, it's not bad." Key measures of quality, though taken, are buried somewhere, which is the opposite of keeping problems visible.

From an upper management viewpoint, a few measures of concern are better. Data dumps do not focus on anything, and if key measures are well chosen, detailed measures at various stages of the operation can be related to them.

Take a broad view of quality. Errors in communicating with customers are a quality problem. Malfunctions of equipment are a quality problem. Poorly executed engineering changes are a quality problem. Quality is performance—a total company activity that cannot be policed into existence by a quality assurance department. *From total quality in operations, quality products and service come naturally.*

Set targets for improvement: goals and target dates. No matter how good or bad the current quality measure is, set a target for improvement. If 100 percent of units must be reworked, set a target to eliminate, say, half the rework in three months. If a defect rate goes down to two ppm (parts per million), set a target of one ppm for next year.

If no targets are set, people rest on their reputations or only react to trouble. Even one justified complaint is too many; the unjustified ones still indicate displeased customers and deserve inquiry, which may lead to improvement in product or in customer perception, education, or service. From such an attitude is a quality operation made.

Consciousness raising is a start, but permanent progress comes from developing people at all levels to attack quality problems. Many quality problems are "old dog" issues, so the development process takes time as well as training, just as in upgrading proficiency in a sport. A golfer does not develop to full potential in six weeks of lessons by using one easily mastered technique. All the attitudes and techniques work in combination, and then only with practice. Trace problems to their roots and eliminate them. Simple techniques of cause-and-effect analysis are quickly taught to non-statistically oriented people: brainstorming, asking the five whys, Pareto charting (categorizing problems by largest category first), scatter diagramming, and so forth. Actual use of simple methods takes time and practice; advanced ones take more. Ask people to think about quality while they work, to look for solutions rather than resign to forces seemingly beyond control.

The root of a quality problem can be anywhere—procedures, design, data accuracy, communicating with the customer, and on and on—so quality is much more than just conforming to design requirements in production. Understanding this is vital to everyone—another reason why total quality is a total company activity.

Responsibility at the source is important. A few years back, Kawasaki U.S.A. began assembling motorcycles in Lincoln, Nebraska, with untrained assemblers. Assembly errors necessitated a large number of repairs by mechanics in an always crammed rework bay. Assembly skills did not improve as expected.

One day, the plant manager (an American) told the assemblers and their supervisor that he wanted to bring customers to the end of the assembly line, give them a motorcycle with the assemblers' names on it, and let them ride it away unchecked and untested with perfect satisfaction. He asked the assemblers to pretend that the customer was right there depending on them.

Then he asked the assemblers to hold up their hand for help when needed and to stop the line if necessary. Rather than do rework, two mechanics roved the line, assisting and instructing.

The following month, the rework rate dropped 80 percent, and the department made full production. Fewer mechanics were needed. Of the remaining 20 percent rework, many of the causes did not originate in assembly, and so management began to work on the fabricators, both in-house and outside suppliers, using the same kind of approach.

Basic as this example is, it illustrates two fundamentals of responsibility at the source: motivation and method. The assemblers became convinced that *their* quality of work was important. Someone else would not correct their problems. A method of improvement appropriate to their level of development was available. These fundamentals apply to more experienced workers too, but the people aspect is more difficult if they have bad habits to unlearn and a reservoir of bad attitudes to empty. Most difficult are staff and professional people who somehow believe that by developing workers, their own competence is being questioned.

Standardization maintains good practice and prevents problems from recurring. Well done standardization is not strict discipline of the unknowing masses but good communication of the whys and hows of performance. It is the basis for broad progress in quality improvement and reduction of waste in general. For a lesson in the value of standardization, compare a McDonald's operation anywhere in the world with those of a franchise that "can't get its act together." Whether a Big Mac appeals to personal taste is not the issue. Consistency is, and consistency is a major foundation of good quality. Few operations last long when consistently bad, but an everybody-for-themselves attitude results in unwanted deviations from expected good performance.

Reduction of variance in the process is a major objective of standardization and also of *statistical process control* (*SPC*), which is becoming the preferred term for *statistical quality control* (*SQC*). SQC carries a connotation of only using statistical measurements, while SPC suggests reducing unwanted variations in the process.

Failsafe operations. Design them so that making an error is virtually impossible or, if that is not feasible, so that errors are immediately detected. Simple, inexpensive failsafing greatly improves quality at little expense. Failsafe design of every activity—from taking customer orders to packing the box—does much to attain near-perfection in quality.

Immediate feedback is the communication of errors, problems, or irregular conditions to the source as rapidly as possible. Poor quality should not be allowed to continue, so immediate feedback is the responsibility of everyone in an organization. The methods are as simple as one worker talking to another or as technical as automatic inspection as each part is made. Immediate feedback is a major motivator for short lead times. The earlier a problem is found, the more easily it is corrected. This principle applies just as much to material going to customers as to a further step in production.

JUST-IN-TIME MANUFACTURING

The phrase *just-in-time* was intended by its originators to imply more than mere delivery just prior to use. The emphasis was on *just,* meaning only what is needed, nothing more. A short definition is "to have *only* the right part in the right place at the right time." Expand the idea to include the total operation from raw material to customer:

Only the right materials, parts, and products,
In the right place,
At the right time.

The operations of directly servicing the customer are included in this definition. Eventually, business-as-usual thinking about marketing is revised just as much as for production. The ideas start small and detailed, but if pursued, end big and broad.

Begin with the relationship between inventory and time. *Inventory* is work stored for future use. It is an asset on the balance sheet as long as it has a future use; but in an operations sense, inventory represents time remaining until action must take place, probably replenishment. If used at a constant rate, a stack of inventory is like a clock showing how much time is left. Some see it more as the length of fuse left on a bomb.

If the lead time to replenish is shortened, not so much inventory is required. If only the right amount of material is in the right place at the right time, almost no inventory is required.

This roundabout logic turns to the lead time for work necessary to replenish. Why does work take time? Then what does work consist of? The issue of waste comes in by the back door. Separate work into necessary and unnecessary activity. Decrease the unnecessary activity and lead times decrease; and if lead times decrease, inventory can decrease. Thus the root of the matter is waste. Excess inventory represents waste—not just bumbling, stumbling, dalliance, but the seven wastes of Table 2–1, the seven sins of which the even well-intentioned are guilty.

Cutting inventory cuts lead times until action must occur. If inventories represent time well in excess of replenishment lead time, they can be cut with little increased tension for immediate action, but at some point inventory cuts apply pressure to do it right the first time. Just the right material at the right time means doing just the right thing at the right time and no more. If neither management nor workers are prepared to deal with that, stress becomes unbearable.

Suppose a machine breaks down. One of the first likely actions is to count the parts from the machine to see how long the plant has to repair it before downstream work or shipments are affected. Once the crisis is over, improve the machine so that future breakdowns are less likely. Progress toward just-in-time manufacturing is progress in crisis reduction. If inventories are reduced, people are "allowed" fewer breakdowns and less time to repair them.

Elimination of waste is not trade-off thinking. That is, do not consider it necessary to increase cost in order to decrease defects, to buy more expensive equipment to decrease downtime, or even to increase inventory to increase customer service. Trade-offs are sometimes necessary, but first and foremost seek and destroy waste.

This contradicts a great body of established management thinking. *Economic* inventory quantities, *acceptable* quality levels (acceptable defect rates), *optimum* plant investments, and *optimal* cost allocations are all concepts that accept constraints. The real objective is to break any constraint possible. Constraints are challenges to identify and eliminate waste.

Workplace organization is a good start in fighting waste. Physical action is concentrated in the plants, but properly done, the waste reduction spreads into every corner of the organization.

First remove every unnecessary thing—*everything*, not just inventory: backup tooling, backup equipment, extra gauges. This is much more than a cosmetic clean up. Pay attention to the reasons why these items are there: quality issues, design problems, maintenance, schedule stability, paperwork snarls. This simplification is the opening gun in the war on waste, so excess items will not all be removed in a week.

Then, the workers should finish detailed layouts or assign detailed locations for everything, which sounds innocent but in most plants is a revolutionary step. The workers take responsibility for their jobs and workplace. They do not do it in one easy lesson—not for them, not for management. It is a change in a way of life.

Create visibility of problems. Visibility comes through standardization so that any abnormal situation is obvious. Again, the problem is that neither managers nor workers enjoy having everything be obvious.

Wherever possible, revise plant layouts for straight through standard flows of material. Not all plants can do this, but even in custom-design production, workplace organization and layout improvement can take place.

If standard flows of material become possible, some version of a just-in-time "pull system" can begin. This system complements workplace organization. It makes setting strict limits on work-in-process inventory simpler. It increases visibility. It is a driver of the waste-elimination machinery. But it has minimal effect unless "machinery" is put in place to drive.

Much of the machinery comes from total quality. More comes from measures taken to reduce lead times, such as reduction of setup times. Setup time is downtime to change from one part to another.

The purpose of setup time reduction is not just to reduce lot sizes. If people setup frequently, they must learn how to do it well—make it a standard, routine operation, not an unwelcome struggle combined with patchwork maintenance while under the gun to get running again. The quality has to be good, specifications known, tooling in condition, equipment well maintained, and everyone knowledgeable about their work.

Reduced space requirements both increase visibility and decrease the distances for moving material. The benefits greatly exceed the value of the real estate saved.

Whenever possible, schedule work in a repeating pattern, which helps establish routine times for setups and other activities. Everyone must occasionally break a schedule for a customer, but waste is minimized by stability and repetition—and by never scheduling to capacity. Extra capacity is people and equipment. They can be improved during off times. Inventory cannot.

Creating conditions for such a schedule is more than technical planning. It takes teamwork from managers and staff with a broad appreciation for the problems of the total company, one of the most difficult aspects of just-in-time manufacturing.

Just-in-time manufacturing makes a difference in all parts of the company: marketing, distribution, finance, accounting, personnel. The changes cannot be confined to production and engineering. For instance, ideas of *how* to serve the customer can radically change if increasing quality, decreasing costs, and reducing lead times simultaneously become possible.

TOTAL PEOPLE INVOLVEMENT

People make everything else happen, whether in an old labor-intensive plant or in one as highly automated as General Electric's Appliance Park dishwasher plant. A people problem, especially a unionized one, is sometimes thought to disappear if humans are replaced by automation. However, a machine is only a tool, even if it is programmed with artificial intelligence.

People problems are not confined to labor relations. Management problems are the source of many work force problems. Progressive manufacturers speak of replacing adversarial relationships with cooperative ones, but personal differences and conflicting ambitions will not disappear. The tough part is coming face-to-face with waste in the mirror, and seeing that what *I* do, proficient as it may be, is unnecessary. The human problems never end, but in the context of value-added manufacturing, a few major human issues are:

Broad Perspective. Specialists may have difficulty communicating—a Tower-of-Babel effect. The stories about narrow-viewed engineers are legion, but accountants and others compile their shares.

Value-added manufacturing is integration of effort. Specialists may not be able to comprehend it, not because they are intellectually impoverished, but because whenever concepts that do not seem pertinent to their specialty are discussed, they lose interest in "someone else's business."

At the extreme, goals of the company are transmuted by specialists into goals of their specialty. An accountant may feel that the purpose of the organization is to make the budget figures work out. A manufacturing engineer may feel that if only more automated equipment were linked together, problems would disappear. A general manager attempting integration of strong-willed specialists blows into the wind.

Every specialty in manufacturing is integrative. That lesson is hard for some to learn. Japanese manufacturers are integrators, even more so if they strongly follow the value-added manufacturing philosophy (only a minority do). They move people into different functions to promote mutual understanding. A young design engineer will do a stint in at least two other functions: (1) hands-on work as a production operator and (2) sales and service face-to-face with real live customers.

The purpose of design engineering is to integrate new technology with both the potential customer and the production processes. How this is done greatly affects schedules, which in turn affect almost everything else. Engineers need to sense for themselves how and why this is so in their company, but they have difficulty fully realizing that theirs is an integrative mission.

Other functions have similar problems. Call it "turf," "perspective," or a dozen other names, it is a major reason for Japanese failures to progress in value-added manufacturing. Similar issues defeat the philosophy elsewhere. In fact, they prevent it from being comprehended except as a bag of tools and techniques.

Problem-Solving Atmosphere. Every work area in a company is a laboratory for improvement. Most managements believe they have that already, but reading between the lines of the washroom graffiti may produce a different opinion. Elevating and sustaining the right atmosphere is the key: constant education, motivation, and participation in problem solving for improvement.

Developing people for continuous improvement is much like developing athletic teams for world-class competition. Training to beat the competition is not enough. Combine method, practice, and motivation to perform as nearly as possible to full potential, day after day. It is not always exciting.

Not everyone can do it. In spring 1984, I boarded a plane in Orlando, Florida, beside two young aspirants to professional baseball, both with outstanding talent and outstanding college careers. One was on his way to join a higher-level minor league team, obviously eager to further polish his game. The other had decided that professional baseball was not for him. I asked the second one the difference between professional baseball and college baseball. "About a hundred more games per year," he replied.

Problem solving requires visibility of problems, responsibility at the source, and other practices not always human nature to accept. The necessary policy is that no one is disgraced by having an unsolved problem, but disgraced by hiding it or not working on it. Problem seeing and problem solving with a minimum of back-biting among people is a challenging atmosphere to create. (In a Toyota plant, a serious problem is "not seeing a problem." The idea is to *enjoy* both seeing and solving them and to stimulate others to see and solve them, too.) Creating the atmosphere is management's responsibility.

Employment Security. The new wisdom implies that a company guarantee employment to at least a select cadre of people. However, no one does that unconditionally, and a major motivational trap for a company commencing huge productivity improvement is the doubtful future of potentially displaced workers. The problem exists for employees in all positions, including management and staff, not just production workers. The dilemma is promoting participation in problem solving with people who understand that their jobs could thereby evaporate.

One can always hope that market share can be stolen from competition so this situation does not become reality. While possible, the hope is also an excuse to duck the issue. There is no valor in that. Prepare an honorable exit plan in advance.

Thus value-added manufacturing requires a survival mentality on the part of top management and its acceptance on the part of employees. Properly prepared veteran employees may work to hasten their own departures. It may not be pleasant, but neither are death and taxes. One example of this is Cummins Engine Company, in which employees are rewarded for figuring out how to eliminate their own jobs so as to take early retirement.

Performance Measurement. Performance measures represent management's expectations and reflect their philosophy of how people work. Some performance measures assume that people do not have to work together very much or very well.

Simple sales quotas do this. Efficiency measures do. Output-based incentives do. They tell people to just do *their* job and to worry little about long-term improvement, which is *someone else's* worry.

The goals of value-added manufacturing are *continuous improvement in* elimination of waste, reduction of lead times, increased quality, reduction in total cost, and people development.

Performance measurement should therefore measure improvement on these dimensions. Not just acceptability, but *improvement*. Not routine performance, but *improvement* over the previous performance. Measure improvement trends and allow for such circumstances as new product introduction.

Performance measurements are the emblems of a management philosophy because people measure what they consider important. When the philosophy of management changes, the measurement systems change—or should change. However, changing measurement systems is more difficult than reworking a machine. Performance measurement is the basis of every system in a company: cost systems, planning systems, capital budgeting systems, personnel assignments, promotions, reorganizations, budget allocations—the mechanisms, built up over years, by which everything *runs*.

A truly accepted change in management philosophy knocks the props from under all this. Some of the most common roadblocks to progress are systems created to assist management in a previous incarnation and to which allegiance and deference are still committed. Major overhauls bring out the same emotions as if the perpetrators were to hold a rock concert in a cemetery. Performance measurement changes are only possible with strong leadership at the top of the company—and those leaders have to be careful if their performance is judged by a horde of impatient investors.

GETTING STARTED

Just start. Most other revolutions are not elaborately planned, either. Form a core group of people who can make a difference and work through the concepts mentally until ideas begin popping in the specific context of your company. Start a few pilot operations. It takes time for people to come around to the thinking, and many of them must see and experience it to accept it, so develop something for them to see and experience.

Start with your own operations. Work with suppliers or others as necessary to improve internally. The whole philosophy is much bigger than just a delivery system from suppliers, but people will jump at that as something that can be done with little delay—perhaps creating more waste than they save.

If upper managers have even mild interest, a few pilot projects should show some visible result after a few months. How long depends on how much the changes become tangled in the decision systems usually used in large companies. Pilot projects should be low-gear nudges to corporate momentum, but many companies stall here. The corporate culture sticks in neutral.

This is the time for upper management to assume leadership. Not all the necessary measures are necessarily enjoyable, and the experience of top managements is paradoxical—an authoritarian push to a generally more participative form of management.

The typical consequence of more responsibility at the source is stronger line management. The staff roles move toward advice and training, though not exclusively. Staff people who perceive changes in terms of winners and losers may not appreciate that.

At bottom, value-added manufacturing is a philosophy of competitive advantage through development of people. In complex manufacturing, the best advantage comes from what people can do that competitors cannot. It is not quickly established. Once firmly in place, it is not quickly destroyed, either.

CHAPTER THREE

Total Quality—A Matter of Detail

Every work activity has a quality aspect, whether customers see and judge the work directly or whether it is part of the background customers neither see nor understand; therefore, everyone should be concerned about quality. No one can disagree with such platitudes, and even though they may not have a succinct definition of quality in a particular case, everyone knows it is often crowded out by emphasis on other aspects.

Emphasis on quality fundamentals becomes tedious, but efforts toward excellence must start here. Failure to pursue quality in the rush to accomplish things that seem more interesting is a major reason for stagnation in the drive toward manufacturing excellence.

The dictionary definition of *quality* helps little, starting with, "Anything which makes or helps to make anything such as it is." That leaves much room for subjectivity, as when interpreting an abstract painting, and yet in manufacturing the presentation of quality depends on taking the right action, which is usually possible only when good quality can be clearly distinguished from bad.

Definitions of quality crafted for manufacturing, simple as they are, require reflection to be fully appreciated. Some of them are:

1. Suitable for use by form, fit, or function.
   a. Form: Size, configuration, density, appearance, and so on.
   b. Fit: Proper working, interchangeability, consistent geometry, surface properties, and so on.
   c. Function: Item performs satisfactorily when used in the customer's application.

2. *Reliability:* Continuing to function as expected over a reasonable lifetime. More formally, the probability of correct functioning during a specific period of time.
3. *Consistency:* Every unit possessing the same attributes, functions, and performance with little variance between them—no lemons. Service to the same standard each time so that customers are confident about what to expect.

Operational definitions of quality do not always come easily, and customers vary in their perceptions of quality. The same customer—the same inspector—does not always see it the same way each time. "The customer is always right" may articulate an attitude to take with customers, but it cannot be accepted literally by auto repair service managers, for instance.

No product is acceptable to everyone, and customer satisfaction is a composite of many factors. Thus, despite the need for precise definitions of quality for operating purposes, the first need is to interpret the requirements of a specific type of customer so that a product can be designed and built to satisfy them. That in itself is a quality performance.

A production definition of quality is conformance to standard or specification. If a product conforms, production has done its job, or so it is thought. The design specifications presumably represent something that satisfies customer needs—which is not always the case. This definition is increasingly recognized as inadequate because all functions are responsible for satisfying the customer, and in practice "conformance" is sometimes argued as any output barely meeting the specification—most of the time, but perhaps not always. Some of the specifications may be designed overly tight with the foreknowledge that they will not be held in production, but the result should be useable anyway. Then conformity is enforced like the tax laws—with serious attention reserved for the flagrant cases.

Because quality is commonly perceived as gold plating and added features, common sense dictates that good quality should cost more than poor. However, quality is a property of plain models as well as deluxe, and basic use at low cost is part of customer satisfaction with plain ones. Products can certainly be made more suitable and reliable by expensive means. Consistency in use can be improved by sorting more defects during production and after, which increases cost. The first effort should be to improve quality with no increase in cost—and probably with a decrease due to less waste. Quality improvement comes from increased skill, quickly knowing what to do, and doing it right the first time—value added with minimum waste.

Total Quality: Defining and Meeting Customer Needs with Defect-Free Products and Services. Nothing is omitted from this definition except perhaps legal issues and financial concerns (such as cash management), and even these have a quality aspect to their execution. Total quality is adoption of an approach to the business, a mind set to have better-than-acceptable quality—continuous improvement of it, no matter how good. Success in adapting the value-added manufacturing philosophy to a specific business depends on assimilating the motivations and methods of total quality throughout the company. Develop total quality in all functions of the business, and from quality operations will come quality products and services. The question is, how?

Strive first for satisfied customers, then for pleased ones. Pleased customers believe they receive more than they had a right to expect. Such an enviable state occurs only if total quality is practiced better than competitors attempting the same, and only if operational specifics materialize from the aphorisms. Gain purpose first, then follow through with details.

HOW IT BEGAN: THE BOOM AND THE FIZZLE

Today Japanese products exported to the United States are generally conceded to be of good quality. That has not happened merely by a quirk of national culture. After World War II, Japanese products had a reputation for poor quality, were an embarrassment to them, and Japanese take embarrassment most seriously. At that time they also imported poor-quality materials—the world's junk—but understood that livelihood in a resource-poor island nation depended on taking what they could get and exporting high-quality, high-value-added products.

The Japanese held a very strong survival perspective. They sought the best methods and accepted drastic changes to implement them. Once emblazoned as a precept of Japanese industry, the motivation of survival through quality has persisted. National policies promoted it in every way. That as much as anything explains the trade surpluses Japan now has with the rest of the world. They do not dare let up.

How did they do it? Dr. W. Edwards Deming explains that when he visited Japan in 1950 he was newly frustrated in his attempts to persuade American top managers to understand their responsibility for quality, and he was determined that in Japan he would not experience the same frustration. He found the Japanese already eager to improve quality, and in his words:

> It was at that time I was fortunate enough to meet Mr. Ichiro Ishikawa, who, after three conferences, sent telegrams to 45 men in top management telling them to come and hear me. Well, I did a very poor job, but I explained what management must do, what quality control is from a management standpoint. For example, I told them to improve incoming materials, which means working with vendors as if they were members of your family, and teaching them. I told them they must learn statistical control of quality, but in uniformity. It's a big job.
>
> I drew a diagram to show incoming materials at the left end and customers at the right end. I explained the importance of understanding the customer, of building and designing to meet his needs years in advance, and of doing consumer research, product tests, and product redesigning, all of which would affect incoming materials and the entire cycle. The cycle is one of continual improvements in line with consumer needs.
>
> This was new in Japan. No one had ever listened more intently or took these ideas up more avidly and with greater success in top management. I had never talked to top management in those terms before, but they understood their obligations and wanted additional conferences to learn more. So I came back.
>
> Well, I think I have put some principles on paper that everybody knew but that, in a sense, nobody knew. They had never before been put down on paper. I stated those principles in Japan in the summer of 1950, some for the first time. They're obvious, perhaps, as Newton's Laws of motion are obvious. But, like Newton's Laws, they're not obvious to everyone.[1]

How did that contrast with U.S. experience? Deming and others instructed thousands of people in quality control in the 1940s. Con-

---

[1] "The Roots of Quality Control in Japan," interview with Dr. W. Edwards Deming in *Pacific Basin Quarterly,* Spring/Summer 1985, pp. 1–4.

trol charting began in many companies. It grew and spread—the fad of its day—but it burned out. Again in Deming's words:

> Brilliant applications burned, sputtered, fizzled, and died out. What people did was solve individual problems; they did not create a structure at the management level to carry out their obligations. There was not sufficient appreciation at the management level to spread the methods to other parts of the company.
>
> The man who saw these things first was Dr. Holbrook Working of Stanford. He knew the job management must carry out. He saw it first. We tried, but our efforts were feeble, and the results were zero. We did know how to do it. In our eight-day courses we would ask companies to send their top people, but top people did not come. Some came for one afternoon. You don't learn this in one afternoon. So quality control died out in America.[2]

Deming explains the reason for this well. Americans thought quality control was a bag of techniques. Feel a pain. Apply a technique. Zap. Control charts appeared in many plants, and for a time people checked on the points that went out of control, but no action was taken for permanent correction. That action would have gone to the heart of the management process throughout the company. Soon supervisors and operators saw that the control charts were just another report management wished to see, something with numbers, as if mere plotting would make anything better. Quality "experts" watched charts but made no lasting changes. Those who needed to take action never really understood.

GAINING PURPOSE

On March 8, 1985, the Association for Manufacturing Excellence held a workshop at LaGrange, Georgia, attended by individuals from some of America's biggest and best manufacturers. Discussion turned to progress in using the JIT/TQ concepts. Most had made remarkable improvements in departments or isolated operations. Some had made considerable progress in individual plants. However, the major problem was described as "how to leverage the pockets of improvement to an entire company."

A basic problem remains almost as Dr. Deming and associates saw it in the late 1940s: seeing the concepts as a management reform for the entire company rather than a set of techniques to be grafted

---

[2]Ibid.

here and there onto business as usual. When the techniques (or rather philosophy) require nutrients unavailable from business as usual, the graftings at best remain stunted and may eventually die.

Perhaps managements do not have a survival perspective sufficient to motivate drastic change. Perhaps changes are regarded only as passing fads promoted in the interest of quickly enhancing the organizational status of their promoter. The succeeding fast-track aspirant has a different program, so changes do not jell before another direction is taken. Developing cohesive purpose for total quality is difficult when top management understands its necessity. Without top management understanding, it is impossible.

Deming is famous for saying that only a small fraction of quality problems (say around 15 percent) are within the domain of the production operator to correct; the rest are management problems. But who is "management"? It is all the rest of the organization—all line, all staff, all functions. Each should come to understand in detail how to directly and indirectly contribute to total quality. The roots of quality problems frequently cut across organizational boundaries, burying themselves in numerous staff turfs.

Creating a linkage between overall company direction and the interlinked detail necessary to support it is a challenge. It requires a bent not only for strategy but for the operating detail that makes it practical, just as great generals must have a mind for strategy coupled with knowledge of logistics, weaponry, and troop capabilities.

Organizational development is necessary so that techniques are adapted to specific products, customers, and processes, and everything woven into a whole cloth. If existing practices are bad enough, it may be necessary to shred the old fabric and start anew, but the objective is to connect all the threads, not patch together fragments.

Figure 3–1 presents a set of conceptual linkages to start the thinking. It is a simple framework compared with an intricate project schedule of cross-functional activities for introducing a complex new product. One way to think of total quality is as a new product or process development, companywide and never terminating. It is a cooperative venture for continuous improvement.

Figure 3–2 is a live chart of the quality system actually in use at NOK, Inc., a manufacturer of mechanical and oil seals located in LaGrange, Georgia. Updated from time to time as their quality approach progresses, this chart is used at the production plant level, so it emphasizes quality in production, but the customer still ap-

FIGURE 3–1 One Conceptual Linkage: Quality in Product Development

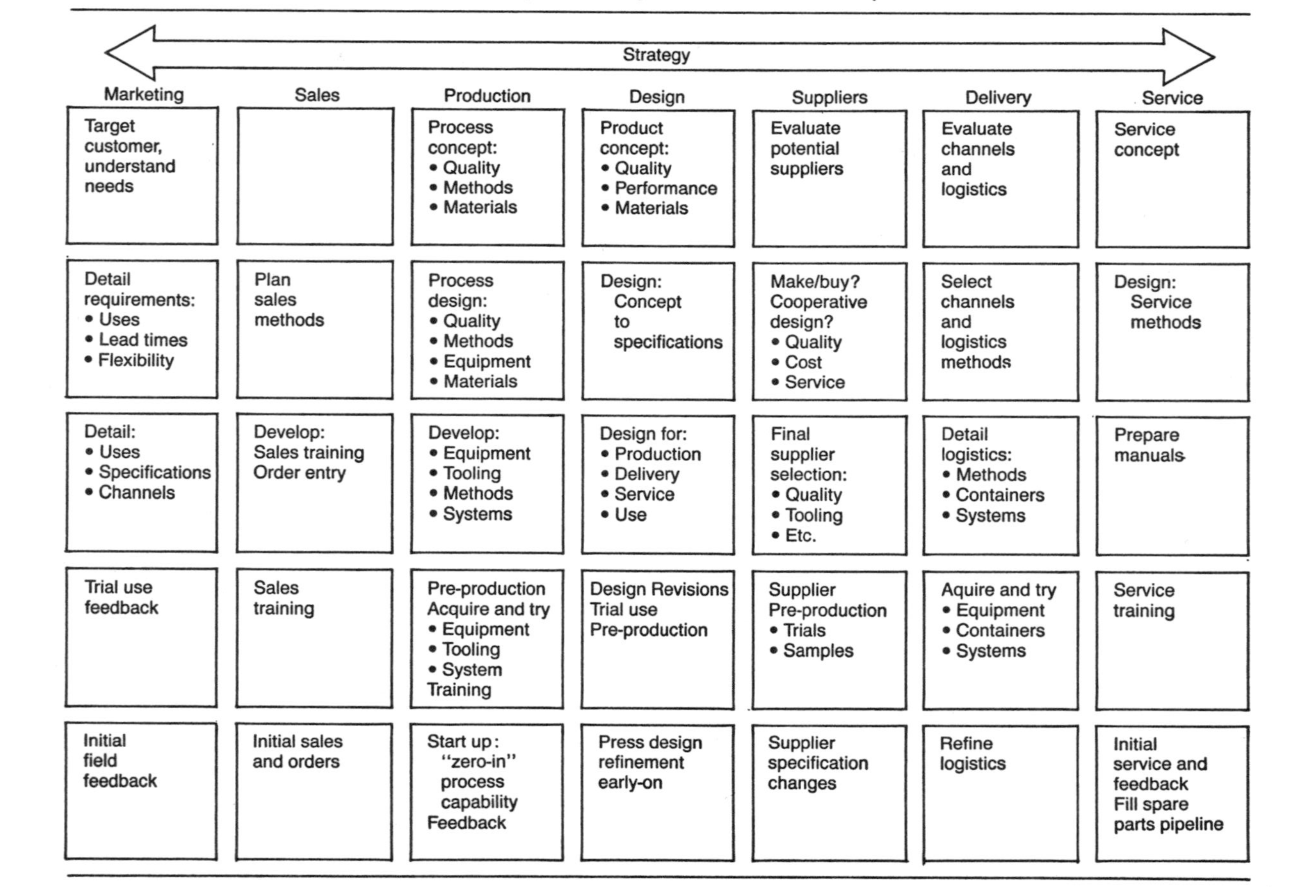

Strategy

Marketing	Sales	Production	Design	Suppliers	Delivery	Service
Target customer, understand needs		Process concept: • Quality • Methods • Materials	Product concept: • Quality • Performance • Materials	Evaluate potential suppliers	Evaluate channels and logistics	Service concept
Detail requirements: • Uses • Lead times • Flexibility	Plan sales methods	Process design: • Quality • Methods • Equipment • Materials	Design: Concept to specifications	Make/buy? Cooperative design? • Quality • Cost • Service	Select channels and logistics methods	Design: Service methods
Detail: • Uses • Specifications • Channels	Develop: Sales training Order entry	Develop: • Equipment • Tooling • Methods • Systems	Design for: • Production • Delivery • Service • Use	Final supplier selection: • Quality • Tooling • Etc.	Detail logistics: • Methods • Containers • Systems	Prepare manuals
Trial use feedback	Sales training	Pre-production Acquire and try • Equipment • Tooling • System Training	Design Revisions Trial use Pre-production	Supplier Pre-production • Trials • Samples	Aquire and try • Equipment • Containers • Systems	Service training
Initial field feedback	Initial sales and orders	Start up: "zero-in" process capability Feedback	Press design refinement early-on	Supplier specification changes	Refine logistics	Initial service and feedback Fill spare parts pipeline

FIGURE 3–2 QA System Chart in NOK, Inc.

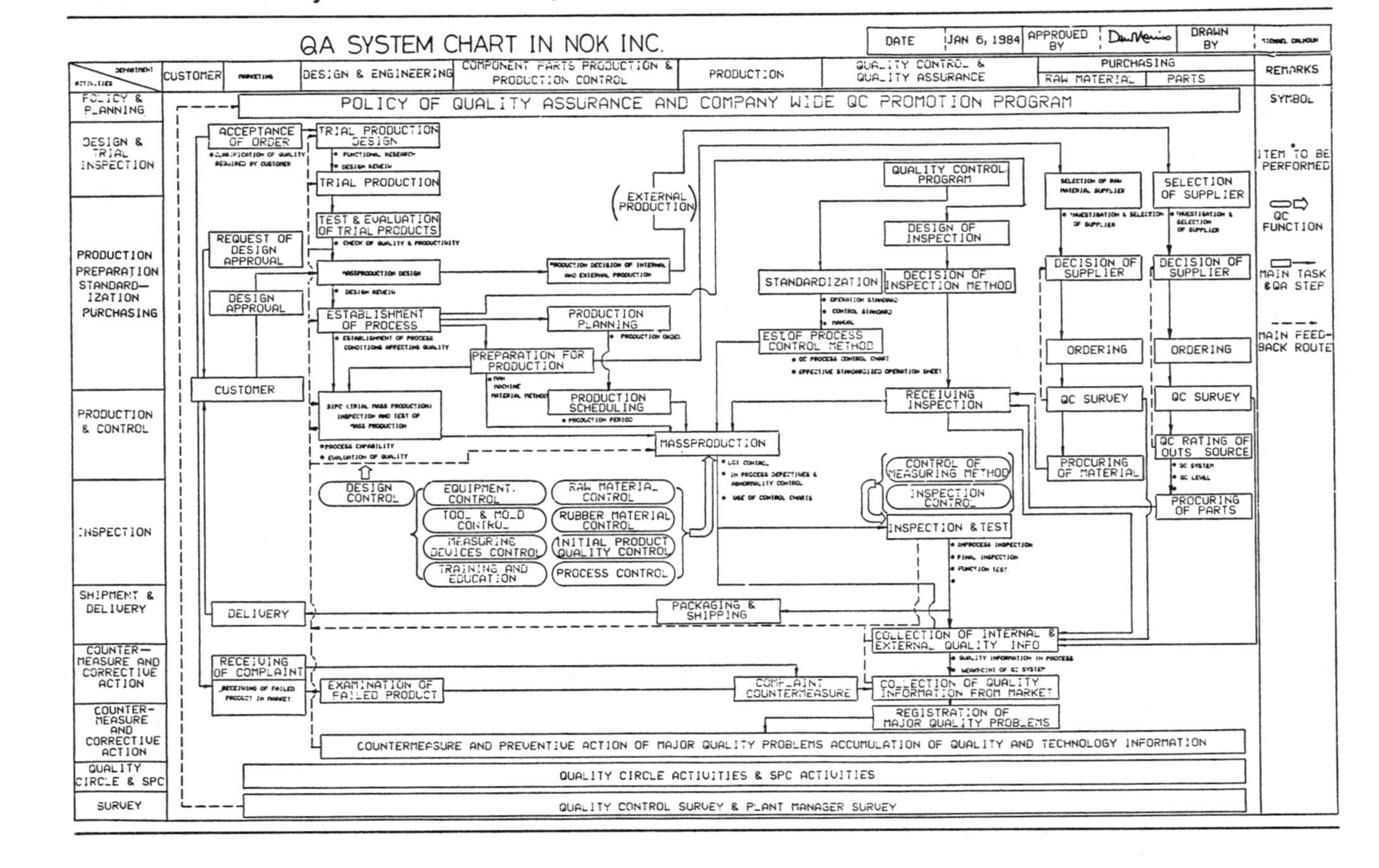

pears on it. Many seals are made for OEM customers, and the seals must meet the approval of the designers of engines or other products.

Note that quality is considered part of everyone's job. Accepting an order is considered part of the quality system—serve a quality customer in a quality way. Production scheduling is part of the quality system: Delay time between mixing and curing the rubber for the seals should be as consistent as possible.

Many quality activities at NOK, Inc., cut across departmental responsibilities. Countermeasure teams, quality circles, and surveys cover it all. NOK, Inc., is a Japanese company, with many Americans in management at LaGrange. They do not call their approach total quality. Quality with the status ascribed to *total quality* need not be designated by a special term.

The company is far from reaching the quality status to which it aspires, but the product is known for its quality anyway. They are busy building it step-by-step into LaGrange operations. Only in their seventh year at LaGrange, these things are not done quickly with a green work force. For example, quality circles just began last year (1985), and they still have only three. Quality circles are created as if cultivating a fine wine.

Both Figures 3–1 and 3–2 are tedious to the point of numbing, but that is their point. They indicate the integrative discipline and the type of leadership required. Gaining an organizational purpose for total quality consists less of dramatic rah-rah than bit-by-bit development at all organizational levels.

Basic education in quality methods is necessary if none has previously existed, or if the emphasis was primarily inspect-it-and-stay-out-of-trouble. Then, set overall, ambitious goals based on the definition of total quality, something like:

1. Understand and define customer needs in detail.
2. Meet customer needs with defect-free products and services.

No one ever meets these goals perfectly, but they set the tone for continuous improvement. Make sure everyone abandons the response mode for quality problems, the condition in which the goal is merely absence of trouble. Emphasize that total quality (or whatever company name is given to the change) is anticipation, prevention, and even more, to make as much improvement as possible. Much effort is necessary to prevent people from being satisfied with lesser goals—and incorrect ones, such as holding complaints to under 1 percent.

Simple as it sounds, begin action by measuring quality.

Translate lofty goals to specific measures. Most companies have measures of some kind, so review these. Do not be surprised to find that some of them are not very useful pointers to problems and are, perhaps, downright silly. If the measures are useful, are managers and workers aware of recent trends in the key ones? If not, asking about quality problems produces the response, "Not bad." In other words, no one has rung an alarm recently.

Track the measures of quality. Are they becoming better or worse? A few key measures should receive top management attention—visibly. A change in the quality atmosphere is stimulated by the mere knowledge that top management is aware of quality measures and expects improvement. Follow up to be sure that the basis of measurement does not change—no performance enhancement by subtle redefinition of what constitutes a defect.

Set targets for improvement. Without targets for improvement, people respond only to alarms. Set targets for improvement when no alarms are ringing, and never accept any defect level as inevitable. Whether defect rates are fifty percent or one ppm (part per million), set targets for improvement.

Measure. Track. Target improvement. This simple one, two, three signals that quality is more important than almost anything else. Now we are going to pay attention to it in detail, but to make quality uppermost in hearts and minds, reinforcement is necessary—especially in time of crisis. For example, Ford Motor Company delayed introduction of their new Taurus and Sable car lines because it was not ready to produce them without fumbling over the problems.

Furthermore, top management must set an example by open recognition of problems, at least among hourly and salaried employees. If management is always proclaiming how great the product is despite common knowledge to the contrary, workers can hardly be blamed for hiding botched work behind a machine in order to repair it later and not show it on a scrap report.

Having problems one has not overcome, and perhaps cannot overcome alone, carries no disgrace. Some problems may be impossible to overcome with the means and resources available. However, not recognizing problems and making an effort is disgraceful.

The converse is the feeling of elation from making progress. Once people begin to develop skill in overcoming quality problems (and others), their morale should receive periodic boosts from the successes. Work should not be an unhappy drag.

If such an atmosphere can be created, top management does not need to set many personal targets. Well-selected targets at the top can be supported by lower-level managers and supervisors who set goals contributing to top-level targets. Figure 3–3 shows a typical top management goal with a small number of contributing subordinate goals. Several features of Figure 3–3 are worth emphasis:

1. It is impossible to verify that all customer complaints are justified, and some complaints are difficult to classify by category. However, the purpose is not to compile a classification on which everyone can agree, but to create a pointer to improvement. "Poor service" could be almost anything: lack of attention from sales personnel, poor purchase recommendation on their part, or the desired item was not obtainable immediately—out of stock. Try to make no-cost or low-cost improvement rather than debate the blame for it all.

2. The intent is not to spend a great deal of money. That is clear.

3. The objective is to improve the technical knowledge of sales personnel, not to increase their training time. Try to make training more effective.

4. By the time items reach a customer, seldom does anyone agree on the causes of handling damage, or even that handling damage exists. Factory, shipper, distributor, and dealer can waste time arguing. Work on containers, methods, markings, and whatever else offers promise of improvement—without adding to distribution expense, if possible.

5. Out-of-stock is not commonly regarded as a quality problem, and it may have many causes, some "justified." Again, the objective is better use of inventory by better allocation, better production scheduling, reduced lead times, and so forth.

6. Customers frequently perceive quality as based more on the nature of the service offered with a product than on the quality of the product itself, as reflected in Figure 3–3. Quality improvement is not just a factory concern.

FIGURE 3–3 Total Quality Improvement Goals

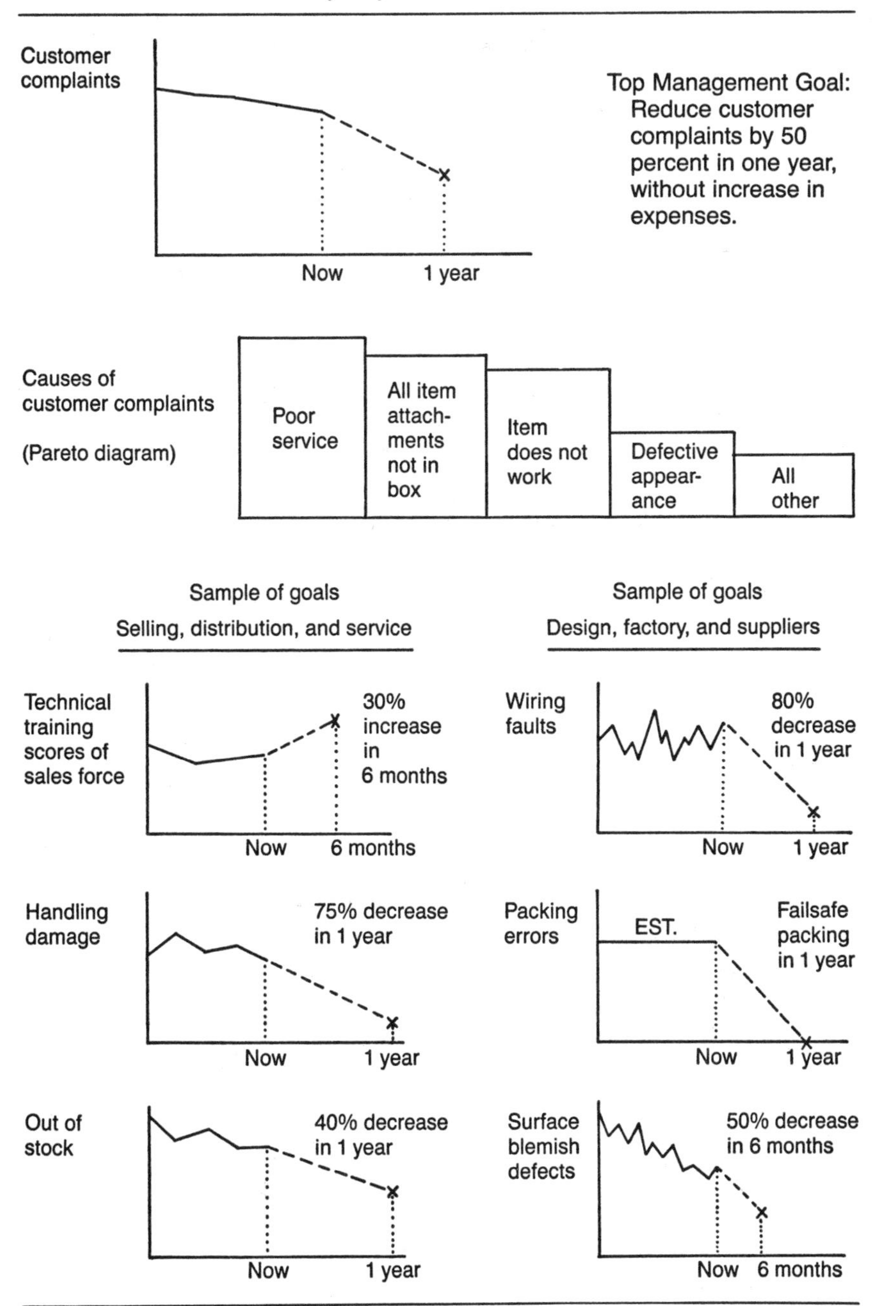

7. Wiring faults can have an incredible number of causes. This goal may lead to a great many subordinate goals and projects, some with suppliers.

8. No one can ever say whether items reported missing from a box opened by a customer were in fact not included when the box was packed at the factory. However, if it is a matter of complaint, try to make it failsafe that all items are packed in each box. Customers may still misplace items after opening or not recognize what they have. Try to make the customer-opening-box process as failsafe as possible, too.

9. Surface blemish is another defect subject to interpretation. The first issue is to establish a standard that everyone agrees is defect free. Then attempt to achieve it. The problems could come from tooling, conditions, material used—any one of several sources or a combination.

Identifying these problems is one matter. Determining how to overcome them with minimum expense is another. The latter comes from experience, ingenuity, and understanding some of the methods used. Understanding only begins with books. The rest comes from working.

Figure 3–4 describes a measure of product reliability. The top management service repair target of 40 percent reduction may be overly ambitious if the work force is not already well developed in quality improvement methods. Several aspects of this figure are not self-explanatory:

1. The subordinate goals are mostly design and production oriented. Examples for marketing, service, and so on are not shown.

2. The relationship of setup and tooling targets to the reduction of dimensional variance are not obvious at first. One source of dimensional variance is differences in setup conditions. If setup conditions can be reproduced within a small variation time after time, the setup time can be reduced because adjustment time should be less. Setup becomes a standardized procedure that is "in control." Therefore, setup times are a good surrogate measure for quality progress.

3. A major factor in setup is the variation in tooling used. If that variation is reduced, both setup time and setup variance should be reduced. Likewise, if tooling refurbishment becomes a more standard, "in control" operation, the time for tooling work should

FIGURE 3–4 Total Quality Goal Setting

No. of repairs per unit in 1st year of service

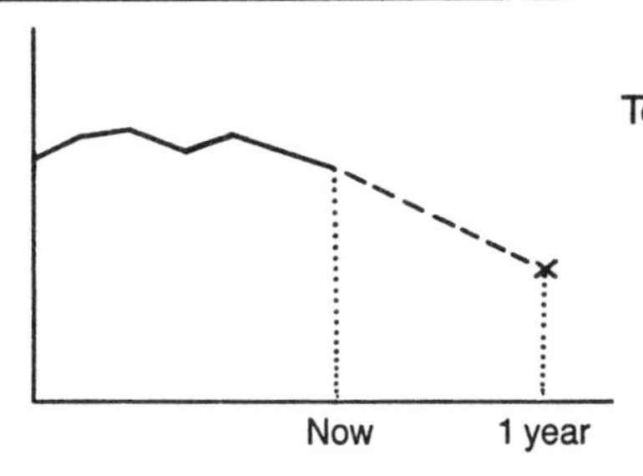

Top Management Goal:
Decrease this measure by 40 percent in one year. (Tough if a year's worth are already in the field.)

Reliability tests
Service reports
Market studies
Improvement efforts

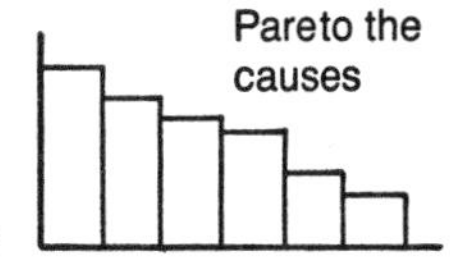

Plant-level goals and goals for sales, design, distribution, service, and so on should complement each other.

Plant-level goals:

Defects after final assembly: ↓80%, 1 year
Total plant scrap: ↓40%, 6 months
Failsafe packing: 1 year
No. of fabrication operations with variance reduced to failsafe level: 25% in 1 year

Motor dielectric defects: ↓ 95%, 1 year.

"Lacquer" process variance in failsafe control: 1 year

Improvement of specifications to supplier and supplier process: "Certifiable" in 6 months

Dimensional variance: "In control": 1 year

Setup variance: In control: 1 year

Setup times: ↓60%: 1 year

Tooling variance: In control: 1 year

Tooling turnaround time: ↓50%, 6 months

Machine availability: ↑50%, 6 months

Non-routine maintenance: ↓ 50%, 1 year

Operator preventive maintenance:

On 75% of equipment: 6 months

On 100% of equipment: 1 year

be reduced. Tooling turnaround time is a surrogate measure for that.

4. Reducing nonroutine maintenance and increasing preventive maintenance should help reduce dimensional variance of workpieces coming from machines as well as contribute to machine availability.

5. Machine availability indicates that equipment (and perhaps tooling) is ready to perform when needed. With few equipment substitutions and delays, setup and run should proceed with minimum disruption, which helps maintain a narrow dimensional variance and preserve quality in other ways as well.

Machine availability and many of the other targets are not in accord with the performance measures used in many manufacturing organizations. Machine utilization is much more common than machine availability in assessing how well equipment is being used, but if its importance is hammered on supervisors, their thinking will be to run equipment as much as possible—"run it into the ground." From the viewpoint of top management, one of the most important aspects of total quality is a feel for performance measures that stimulate it. Much frustration at the working level stems from performance measures penalizing the activity necessary for quality improvement. When establishing performance measures, top management should understand the activity they are trying to promote. Consistency in performance measurement is one of the most important steps in gaining purpose.

METHODS FOR IMPROVING QUALITY

The objective is to improve quality. Merely measuring quality does not improve it, but a major reason for measuring, by simple means or sophisticated, is to determine *how* to improve quality. An organization doing this constantly needs a systematic working methodology, something like the one given here in six steps:

1. *Plan improvement.*
    a. Measure and track quality status.
    b. Set specific targets for improvement.
2. *Systematic search for problems.* Are we working on the right problem, or something that makes little difference? Look in

the easiest places first. A typical routine for locating a defect in a subassembly, for instance, would be:

a. Check for *location* in subassembly. Which component(s)?

b. Check for *time* of occurrence. Regular or random times?

c. Check for *operation* where defect occurs. Is it consistently traceable to same machine, same tool, same operator, same maintenance routine?

3. *Systematic search for causes*. Ask why five times, draw a fishbone diagram (or causal tree)—in other words, sift the evidence for a conclusive sign of cause and effect. In more difficult cases, this may be a long-term experiment, holding all factors constant except the one being investigated.
4. *Propose corrective action. Try it.*
5. *Verify that corrective action is effective.*
6. *Standardize into permanent practice*. The adherence to standards thus established provides a basis for further improvement.

A four-step outline with very similar logic is known as the Deming Circle: Plan, Do, Check, Action. Devise a systematic approach appropriate for the organization based on the scientific method.

This seems a bit underwhelming to most professionals: no great breakthroughs, only careful attention to detail. A full-blown, by-the-numbers systematic approach seems justifiable for NASA engineers analyzing the explosion of the Challenger, but not for operators studying, say, graininess in carrot bags. Indeed, many problems are quickly overcome without elaborate methodology, many by intuition—"logic" not formally stated in any way and sometimes not understood by its users.

The role of intuition is not cut out by the systematic approach, but solutions are difficult to standardize unless they are more clear than intuition often makes them. Problems arise when countermeasures derived from intuition are not checked and documented. Typical is the ingenious but undocumented creation of software. After a time, its maintenance or enhancement may be difficult for the creator and impossible for others. Without systematic organization, the job is incomplete.

Some product or process characteristics are impossible to define without a "feel" for them, for instance, the free-spirit rumble of a Harley-Davidson engine or the texture of a poromeric (artificial leather). The attractiveness of the texture itself is art. Reproducing it time after time is science.

Persuasion is another factor that is not totally scientific. A well-tested, logical solution to a quality problem is of little value if the implementors are not persuaded of its value. Nothing works well in a factory, among a sales force, or with suppliers unless the people doing the work understand it and are persuaded of it. They need extensive training and experience with quality thinking. The power of the systematic approach is the development of many people at all levels to attack problems in a coordinated way—with a shared understanding of cause and effect and how progress is made. The difference between a few people understanding total quality very well and shared understanding throughout a large organization is time-consuming development of people.

STATISTICAL METHODS

Many quality issues are described using means, variances, process distributions, random sampling, and measurement errors. Knowledge of basic statistics helps in understanding some types of quality problems but is no guarantee of overcoming them. This book is limited to a few statistical illustrations. Statistical education is available from many sources but should be presented so as to apply to quality issues.

The heart of "statistical thinking" is understanding variance and random variables. For instance, a control chart can be used to infer whether samples of plastic parts from a molding process exhibit a pattern of mean weights indicating a normally varying operation or whether an "abnormal" condition disturbs it. Few people learn how to do this by an extension of common sense.

Untutored people make all manner of erroneous judgments about variance and random variables. Las Vegas and Atlantic City provide constant examples. They also provide examples of people skilled in manipulating statistical symbols whose judgment at a gaming table seems little improved by the skill. They understand statistics, but not the total context of the game.

A few hours of tutoring begins to sensitize people to statistical concepts. If they are already expert in their processes and begin to apply what they know right away, they may begin improving quickly; usually, however, progress is not so rapid. They must couple statistical concepts with process knowledge and ability to actually make improvement. This takes much longer than only learning to fill out a chart. Even if charts are well interpreted, follow-up improvement activity may not be organized. For these reasons, the time to make quality improvement is often underestimated.

Fortunately, many statistics used to identify quality problems are descriptive and do not require much interpretation of whether or not a process is following a normal statistical distribution. Simply classifying defects by category stimulates the classifier to ask why each defect occurred, which may lead to ideas for preventing them. Such descriptive statistics are seldom difficult to understand. Selected descriptive measurements can lead to much improvement without the mental effort required to appreciate statistical distributions. Even this takes training, not because understanding the measurements is hard but because the habit of logical, causative inquiry in conjunction with those measurements must be inculcated.

The most common descriptive method is the Pareto diagram (shown in Figure 3–3), a simple classification of defects, complaints, or problems by category. The hypothesis is the *Pareto rule*—80 percent of the problems come from 20 percent of the sources, so the most productive way to attack defects is to attack the cause of the 80 percent. Focus effort on a few problems at once, which comes naturally.

By contrast, the pursuit of random variables requires training in the nature of random processes. Control charts are not as natural to create and interpret. Great misunderstanding is likely to arise in attempting to use them. As shown in Figure 3–5, a control chart is used to interpret whether a process is following a statistically random pattern. The figure assumes that samples of some numerical measurement (for example, weights of a plastic molding) are taken from the same process. *Same process* means same machine, same mold, same resin, same conditions: that is, set up as nearly as possible in the same way. The result is the natural oscillation of the production process.

Measurements are taken of the process as it is, not as one would like it to be. The upper and lower heavy lines are control limits derived from observation of the process itself, *not* taken from tolerances or specifications one might like to attain. The process in-

FIGURE 3–5 Statistical Process Control

If the process is in control, the dots representing sample means should fall within the shaded regions with the frequencies shown at right. Less than 0.3% should fall outside the upper or lower control limits (heavy lines).

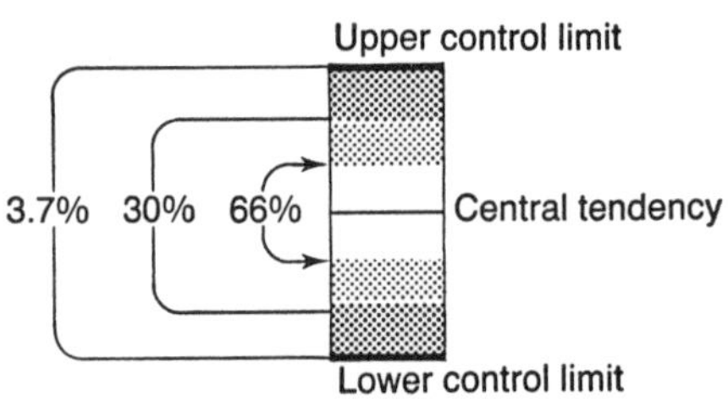

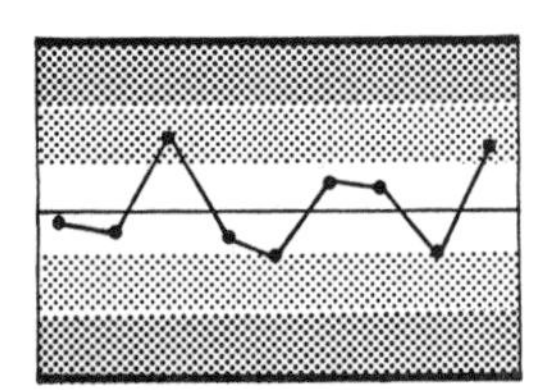

Normal or "in control." Only 2 of 9 samples outside the 66% region, but occasional extremes are possible.

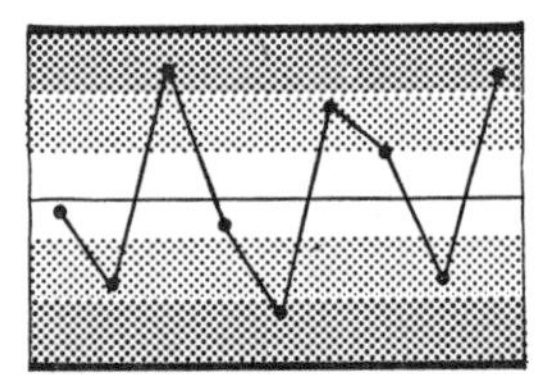

Too much variance. Three of 9 sample means are in the 3.7% zone; only 3 in the 66% zone. Very unlikely if process is "normal."

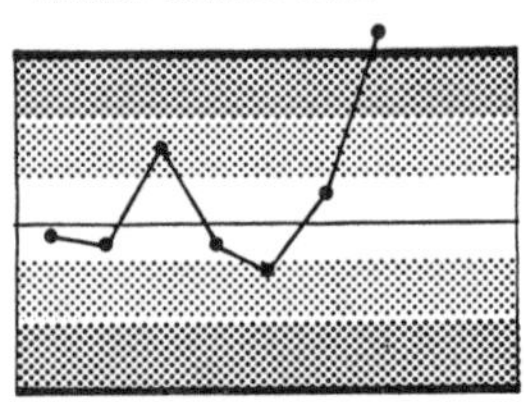

"Out of control." Even one sample outside all three zones is reason to suspect a cause, although rare measurements will normally do that.

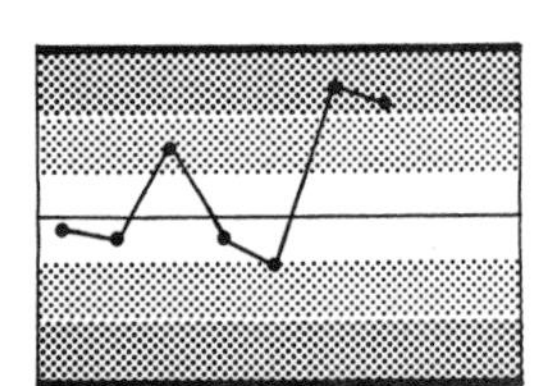

"Out of control." Two consecutive samples are high in the 3.7% zone. Very low probability of that unless process has changed.

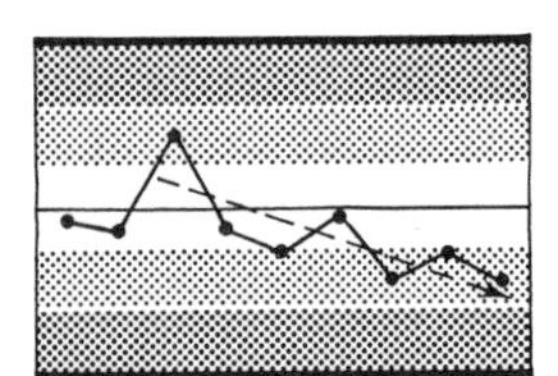

A downward trend is detected. Look for the cause of this "drift."

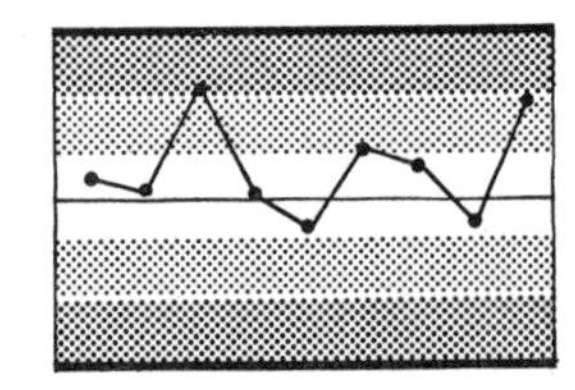

The variance is normal, but the central tendency is too high. Seven of 9 samples are above the historical central tendency.

When A Process Goes Out Of Control, Begin Diagnostics To Determine The Cause And Take Corrective Action If A Cause Is Found.

cludes every variable that might influence the results: machine, tool, material, operator, methods, and adjustments.

A control chart in no way guarantees quality results. In a sense, it merely quantifies the state of ignorance because the normal variations of a process come from causes that may not be readily identifiable. As shown in Figure 3–5, one use of a control chart is to determine when a production process (or any other process measurable in a similar way) has digressed from a pattern previously observed as normal to it. For the control chart to be useful, this digression must be detected for what it is, and the cause found and eliminated if possible. In other words, deviant behavior from control chart measurements should signal corrective action if anyone knows what to do. Therefore, effective use depends on recognizing onset of a nonrandom pattern, knowledge of the process itself so as to diagnose the probable cause, and, most important, the ability to make corrections.

In fact, an operator who does not understand the chart can easily continue a common habit of overcorrecting adjustments. If the chart reading dips, an adjustment corrects upward; if the chart upticks, an adjustment corrects downward. It sounds sensible, but if correction is made for random changes—changes of unknown cause—the operator is chasing a random variable. By constant adjustment, the chart itself does not reflect a constant set of process variables, and the usual result is a wider band of variance than if the process were left to run of its own accord until a definite deviation occurred.

Process Capability

The first statistical objective is to operate a process so that it remains in control. Personnel learn how to run it time after time so that results are reproduceable. Just sampling and charting does nothing by itself.

The next objective is *process capability*. Determine whether the natural variance of the process is capable of producing items that meet the requirement time after time. If defects can be produced just by a random variation, the process is not capable of producing the quality desired; therefore it must be improved by contracting the variance through improvements that change the control chart picture, as shown in Figure 3–6. Figure 3–7 shows some of the

FIGURE 3–6 Reducing Process Variance so that All Parts Are in Tolerance

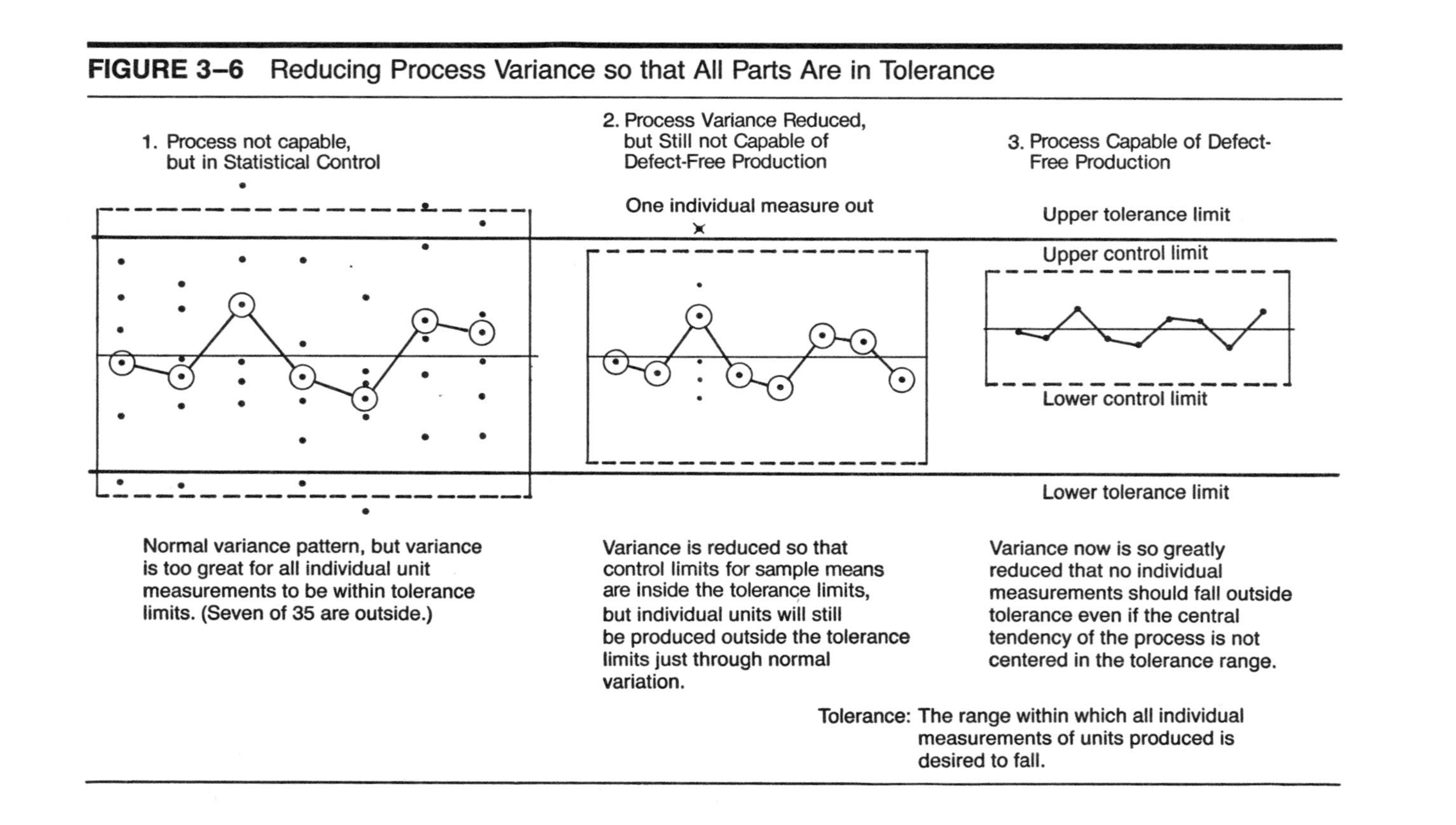

FIGURE 3–7 Process Capability: Sources of Variance

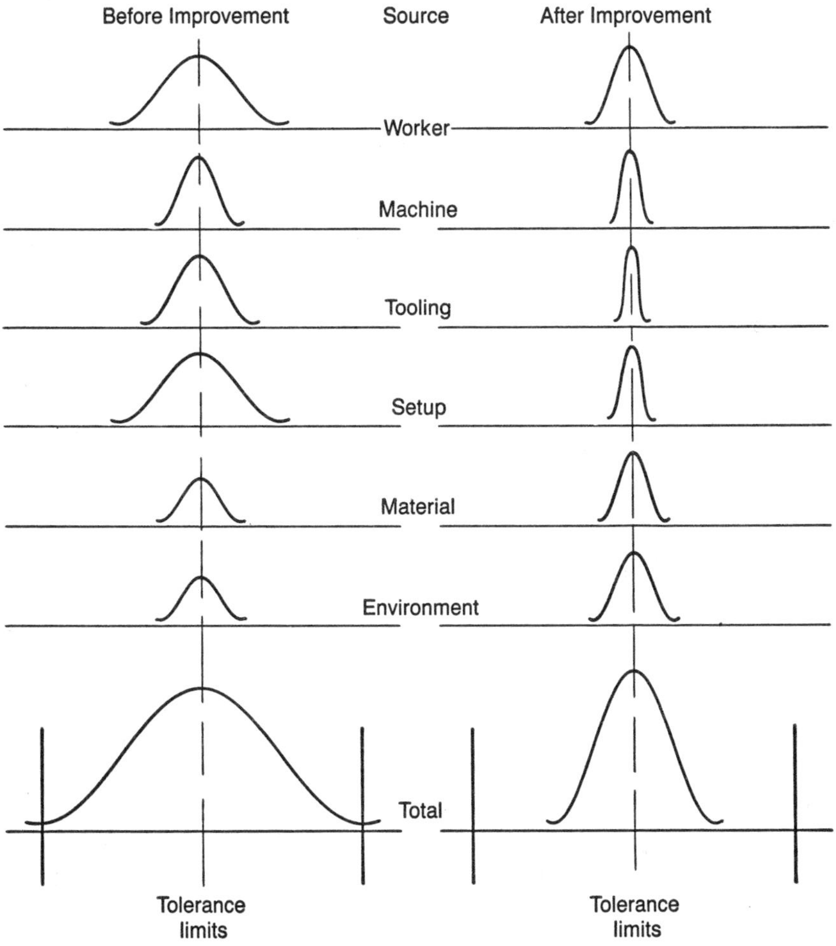

sources of variance. The figures also illustrate a clear difference between the design specifications established for a part and the study of the production process that makes the part.

The design process cannot be independent of the production process or else engineers would ever be calling for specifications that cannot be made without great expense. Sometimes that is necessary, in which case the production process must be developed for it if possible; however, careful design attention can avoid needless waste in specifications that add expense but little value. One company starting total quality found that it had to review 60 percent of all design tolerances. Turnover of design engineers over time had eroded whatever rhyme or reason might have once prevailed in the tolerance stocks assembling the products. Consequently, they were at loss for a reference base when determining what the capability of the process variances needed to be.

Variance is frequently reduced by spending a small amount of money on maintenance replacing worn machine parts, for instance. Observing the statistical effect begins a basis for effective maintenance, but to complete it, the duration of the effect must be assessed. Production processes do not oscillate in narrow-band variance forever. Suppose a particular bearing is discovered to last only 200 hundred hours before wobble is detected from charting. Regularly replace the bearings before degradation starts—for example, at the end of 190 hours. (If this finding is possible without charting, well and good. The objective is improvement, not statistical fussing.)

A policy like this is frequently attacked using trade-off thinking. The argument is: Try to find the optimum point between bearing life and cost of defects. This is false thinking. By extending that idea to every factor that could affect variance in the process, the process is apt to always verge on the loss of control, if not over the brink.

Many of the detailed topics in this book have as a major objective the reduction of total process variance—improvement of *total* process capability: quality cycles, failsafe, immediate feedback, standardization, cycle time concept, cycle time analysis, visibility systems, and on and on. A major objective of properly applied JIT/TQ methods is to reduce total process variance in the broad sense.

Quality Cycles

A better idea is to establish *quality cycles* for all major process factors. For example, a milling head sometimes lasts 30 workpieces on a particular cut before it needs changing, but occasionally surfaces become noticeably rougher after only 20 pieces. Set the machine to automatically stop for a milling head change at the end of 20 pieces. The objection to this practice is that if the wear cycle averaged 25 pieces, tooling cost would "increase 20 percent."

The solution by trade-off reasoning is to examine each workpiece or monitor tool drag until signs of tool wear are obvious. Instead, see what can be done to provide more consistent cutting heads, more homogenous workpiece material (to narrow the variance of cutting time on each workpiece), or other measures to contract the variance in tool wear cycles. Doing this will likely extend tool life anyway, and the outcome might be as startling as consistently seeing deterioration begin in the range of 28 to 30 workpieces. If so, set the machine to stop for a tool change at 27 pieces.

Tool damage is less likely. Equipment damage is less likely. Tool life is actually extended, and equipment life may be also. The operator need not check each piece. Sensing tool drag is unnecessary. (If drag is checked, check for increase in the drive motor current, and shut off automatically if it increases. That safeguards the entire machine.)

By placing tool change on a *regular time cycle,* the methods of automation can be simple. Tool use is on a regular basis, too. Quality cycles contribute to setup procedures, and they may form a basis for lot sizing.

Control Charts Are Unnecessary on All Operations at All Times. For many operations, they may never be necessary. Process capability is established by straightforward observation, and zero defects are virtually assured by simply comparing the first piece from a new setup with the last piece from the previous setup. Most processes will not jump out of control and back in again of their own accord. Procedures to failsafe quality setups and setup changes can be instituted once an operation has been studied and the quality cycles are known.

The purpose of statistical methods is to learn something that might otherwise be unseen, then use the knowledge to improve the quality of manufacturing processes in inexpensive ways. Use them to learn something. A well-developed sensitivity to statistical variations greatly assists discovering ways to improve. Mindless charting does not.

Managers, engineers, and operators gain great insight by understanding that every activity can have statistical variation associated with it. Statistical variance helps in understanding and resolving even the measurement errors in taking the readings. Measurement errors are often the biggest source of variance observed, and a common problem is discrepancies between different instruments used to measure the same thing. Professional attention to this problem is necessary to allow others to attack quality problems.

Using statistical methods to improve process control and capability is coming to be called *statistical process control (SPC).* The term *statistical quality control (SQC)* has been used for many years to mean the same thing, but over time the meaning of SQC has been corrupted because companies labeled as SQC many efforts that only measured the quality of a product with little attempt to improve its design and manufacturing processes.

The easiest version of degenerate SQC to recognize is a program that is only statistically sorting defects—postproduction inspection. One objective is to so improve the process itself that special inspectors can be eliminated.

More subtle forms of self-deceit exist. One is poor application of statistical methods: statistical methods not properly designed, actual measurements not correctly taken, or faulty interpretation of results. Another deceit is using correct statistical approaches but failing to follow up by diagnosing causes of defects and taking corrective action. On-line and real-time statistical measurements do not make improvement by themselves. The habit of good "statistical thinking" takes much time to develop in a work force. When effectively developed, it should lead to failsafe process improvement whenever possible.

Failsafe Methods

Failsafe methods, sometimes called foolproof methods, are a well-known approach in design. The warning buzzer to buckle a seat belt is a mild form of failsafe. Check digits in data processing are another. Automatically stopping the machine at 27 cuts to change milling heads is an example from production. Failsafe thinking is all around us, but the objective is to incorporate failsafe thinking broadly and inexpensively in the total manufacturing process.

About three years ago, Black & Decker tried a failsafe method for packing bench grinders in boxes. The basic consumer tool, plus attachments, instructions, warranty cards, and the like, totaled nine items in all. Before that time, about 70 percent of all customer complaints resulted from all items not being present in the box received.

The grinder was placed in a box on a conveyor and was stopped by a trip arm at the packing station. A set of bins holding each of the eight items added was built, each with a photocell mounted so that when the packing operator's hand reached into the box, a beam of light was broken. Unless all eight photocells were tripped, the trip arm holding the box would not release it to the sealing operation.

This is not perfect failsafe. Just because a light beam is broken is no guarantee that the operator actually grasped an item and conveyed it to the box. However, nine months after this trial started, not a single customer complaint was judged to have originated from incomplete packing. This form of failsafe was simple, inexpensive, and effective. It has now been extended to all products in the Black & Decker line.

Another example of failsafe is a rubber mixing operation at NOK, Inc. A microcomputer is linked to the scales used to weigh each component added to a rubber blend. The computer displays one at a time each component to be added, its bin location, and the amount required, in sequence. Only after each component weight on the scales is attained within a narrow tolerance range will the computer move on to the next item. The consistency of rubber blends has been much improved by this procedure. One of the major problems overcome was a common human error of being distracted in the middle of a weighing sequence and forgetting which component is to be added next.

No method is so failsafe as to be impossible to circumvent, but if desired, more safeguarding is obtainable at even greater expense. Ethical pharmaceutical manufacturers require that all counts of bottles, labels, cartons, and literature be reconciled both before and after the filling process. This procedure is similar to the cash counts in a bank. These procedures are affordable because the users will pay for ultimate failsafe. No error is tolerable.

Manufacturers with low-value-added products cannot afford this because failsafing many possible sources of error at great expense costs too much. Better to develop very inexpensive methods of failsafing. Build them into operations with minimum capital expense and no added labor. Simple failsafe methods greatly reduce chances of error, especially human error, and reduce costs. They are the low-cost route to parts-per-million error rates. Apply them to error-prone operations and to preservation of the quality cycles.

Failsafe methods do not improve basic process capability. In the case of the milling machine, further improvement of machine, milling head, or material may further extend tool life, and then the operation of the machine can be failsafed again at a higher piece count per tool.

Immediate Feedback

This is a basic operating rule: Communicate the existence of a defect or of an irregular condition to the point where corrective action will be taken as quickly and accurately as possible.

Establish zones of quality with another simple rule: Accept no defects and do not knowingly pass on any defects. If defects proliferate, this rule is impossible to follow literally. Nothing would ever ship. Start applying the rule to at least some kinds of defects and keep tightening as people and processes improve.

A zone of quality may be a plant, a department, an operation, or a single worker. In any case, transmit the news of bad quality directly to the source, whoever that may be. That is simple enough, but it violates the chain of command. A worker finding a defective part should send it straight to the worker who made it—no piddling around with the political intermediaries, and no delays while someone ponders how they are going to look to a superior.

For this procedure to work, people must learn to approach it in a problem-solving manner. If one worker, or supervisor, tells another, "Look what you did to me this time," or questions their competence and sanity, emotion will delay corrective effort. The purpose is to immediately fix the problem. Only when this attitude prevails will person-to-person immediate feedback be productive. Developing the attitude is the hard part. Unfortunately, some of the names used to describe this practice have not been helpful. One Korean company said that all employees should become "quality spies," for instance.

Direct point-to-point communication on quality also affects the role staff people see for themselves. Many think the workers should immediately consult an expert—them. It takes no small adapting for staff engineers to learn that a major job should be coaching workers to do routine troubleshooting for themselves. Eliminate conditions in which workers stand idle while a few experts correct even the simplest malfunctions.

The time to find a defect is the moment after it is made. Better, detect an irregular process condition *before* a defect is made. In short, check parts immediately after the making or check the process immediately before the making. Build gauges into production equipment if it can be done simply. Checks for presence of parts or configuration can be built into jigs, for instance. Locate gauging as near to the scene of action as possible—no long walks unless a supersensitive gauge is required. (Among the reasons for lengthy setup times in one press department: workers required at least two hours to walk to a quality lab and wait for parts to be check gauged there. Several trips were sometimes required.)

The sooner an irregular condition is detected, the better. If parts or materials are returned to the scene six months after production, reconstructing what might have happened is much more difficult. Detailed records are expensive and still not as effective as direct observation.

Immediate feedback is aided by four other goals in production:

1. *Short lead times.* The less time after production a problem is found, the better.
2. *Low inventories.* These go in conjunction with short lead times, and if anything goes wrong, less material needs to be examined, sorted, or thrown away.
3. *Frequent setups.* Every setup is a quality check, too. A long run is not allowed to degrade, and run lengths (lot sizes) can be based in part on good quality cycles.
4. *Small lot sizes and small containers.* If a process goes sour, the defects are confined to a small amount. This is especially apparent for a machine making small parts. If it starts to turn out defects, there is a great difference between sorting a small box and sorting a large barrel.

Immediate feedback is also an objective of maintaining a high degree of visibility in operations. If a machine or operator has a problem, a light or other signal notifies everyone within sight. If a JIT operation stops for very long, the small amount of inventory causes other operations to starve out of material, so a hiccup in smooth operations is tended before it turns into a choke. Naturally, the inventory levels should provide time for correction delays appropriate for the degree of development of both people and process, but the intent is for everyone to pay attention to quality problems immediately, not when they get around to it.

Immediate feedback assumes a high level of responsibility at the work station. It comes from the spirit of problem solving—people development, not just system development.

Immediate feedback is an objective with suppliers, too. JIT arrangements with suppliers are incomplete unless they include the development of immediate feedback, which is point-to-point communication with as few intermediaries as possible. "Communicate from the point where the defect was discovered in the customer plant to the point where the defect originated in the supplier plant." Implied: Top priority between the companies is problem resolution, not negotiated returns, fixing of responsibilities, or who-shot-John stories. Fix the problem and complete any necessary business adjustments later.

Cause-and-Effect Logic for Collaborative Problem Solving

Quality will improve if an organization is able to do two things better: (1) identify the true causes of quality problems and (2) take effective action.

Most professionals take pride in their individual problem-solving ability. Most likewise take pride in teamwork to solve problems. Problem solving is recognized as part of everyone's job, but that recognition dims if the job is direct labor. However, few people will disagree that improved skill in collaborating on problems would make a better company and a better product.

A development director at Union Carbide followed this practice 25 years ago. His development group installed new products and processes in polymer film extrusion plants. Most young engineers wanted to turn a new installation over to plant management only after it was demonstrated to be clearly superior. The director was wiser. His approach was to involve production supervisors and operators early and to let them finish debugging, discovering some of the development oversights, which made engineers feel foolish. However, in the end, the production personnel did their best to make a capable process run.

In that bygone era, hardly anyone thought of sharpening the general cause-and-effect logic used by both professional and operative personnel. The outlook was, and still is, that once a new development is running smoothly, the development effort is done. Production will make occasional improvements, but operation is their primary responsibility, not development.

Total quality is a different perspective. Quality improvement is a never-finished development project. In order to be effective, train everyone in some basic common methods for assessing quality problems. It promotes communication. Quality problems can have their roots anywhere from the design concept onward.

There are many cause-and-effect methods. For manufacturing purposes, two of the better ones are Pareto diagrams and cause-and-effect diagrams, because they lend themselves to collaboration in problem solving. Collaborative methods are useful when the root causes may be anywhere or may be collective poor practice. An expert may much more quickly knife through technical solutions. However, many quality problems result from many little oversights with responsibility shared among many people, and their solution depends on recognizing the problem through the organizational and technical fog. Collaborative cause and effect helps many people see the nature of the problems and appreciate the solutions, which is helpful if cooperation is needed in taking corrective action.

The Pareto diagram is so simple a concept that it can be practiced with very little instruction in the technique itself (examples in Figures 3–3 and 3–4). The value of Pareto diagrams is in their construction as well as their interpretation. In order to classify defects (or other phenomena) into categories, the classifiers must ask themselves over and over why a particular defect occurred; then, they can do it again in classifying defects into subcategories and sub-subcategories; finally, some ideas for corrective action may result. The diagram is simple, but methods used to identify defects and determine their causes may not be.

Pareto classifications are only one way to codify and summarize the quests for why. Other diagnostic routines may be better, depending on the situation, but the desired feature of any is to organize data so that many people can contribute to it and to post the results so they are visible to all. Focus many minds on the problem, not just a few. The *habit* of open, collaborative problem solving is hard to promote.

Likewise, the cause-and-effect diagram is a very simple method that can be used in many variations. It is especially useful in situations where the objective is to transfer many observers' insights into a pattern that makes overall sense and that makes all somewhat aware of the state of progress. The cause-and-effect diagram is simple; its use is not. For a much more extensive discussion of cause-and-effect diagrams and their use, see *Managerial Engineering* by Ryuichi Fukuda.[3]

A key quality point is a point in a production process at which the quality of several prior operations can be determined. In electronic manufacturing, the check just after wave solder is a key quality point. (Most other kinds of production have their own key quality points.) The results of many prior processes can be checked there, and the problems discovered may be rooted in any of them. If defects are high in a new process, assigning an expert (probably an engineer) to run down and correct quality problems can considerably decrease defects without using any special cause-and-effect logic at first. The first wave of problems and countermeasures may yield quickly. Moving on down to lower defect target levels becomes more difficult. Collective cause-and-effect approaches are reserved for problems that remain.

Although this discussion of methodology is limited, *developing all employees to systematically see and attack quality problems is the key to success in many areas*. The advantages are twofold: (1) improved skill and awareness on the part of many people and (2) when improvements can be gained just from careful observation of undisturbed production processes, costs are less than if special experiments or tests are run.

Without corrective action, continuing the search for causes has little point, and effective action can often be taken without precise evidence. For instance, if one is trying to reduce the dimensional variance of a process in statistical control, precise determination of how much variance comes from each of a number of sources may take more time than it is worth. For starters, tend to the obvious things, like replacing worn bearings, followed by simple approaches to improve dimensional reproduceability of the setup. Actions that make improvement may help clarify further actions to take and thereby become part of the cause-and-effect logic itself.

Corrective action removes part of the problem and clarifies the remaining problems. Much of total quality consists of doing one's best to start up a low-defect process at the beginning of a new product life cycle and then never ceasing to find ways to make it better. Squeeze the great fuzzy set of unknowns, mysteries, and

---

[3]Ryuichi Fukuda, *Managerial Engineering,* Productivity, Inc., 1983. P.O. Box 16722, Stamford, Conn. 06905.

black art into a smaller and smaller ball. It may never disappear, but make it insignificant.

STANDARDIZATION AS A BASIS FOR IMPROVEMENT

Standardization is used in many contexts in manufacturing and implies any number of ideas, starting with compliance to industry or government codes. Internal standardization includes:

1. Reduction of total numbers of parts and materials used.
2. Sometimes a reduction in total numbers of products, models, or grades produced. Standardization of a product line.
3. Use of the same equipment or same type of equipment.
4. Use of the same kind of tooling in a similar way.
5. Consistency in methods used for material conversion, set-ups, locations, material handling, and so on.
6. Standardization in equipment and methods, which permits standardization in training.

Standardization promotes quality through simplicity and consistency, which is so obvious that sometimes it is overlooked. Standardization has a bad connotation because it restricts everyone's freedom to do their work as they might like. That is part of the intent because not all individual variations improve established methods.

A reduction in total parts assists quality effort because fewer items must be made or purchased, and the fewer there are, the fewer ways for something to go wrong. Costs should decrease also.

Standardization of a product line is not as easily related to quality because the first objective of the product line is to satisfy the customer, and the real purpose is to standardize the process of *providing the product line* without compromising its value to the customer. Perhaps one of the clearest examples is from process-type manufacturing, as for polymeric resins or integrated circuits. Variations in raw material and in the process itself may make predicting the yield of the process impossible, so results are tested and graded, and the output sold in grades. In the case of polymers, the grades might be based on molecular weight distributions. If variance could be removed from the material and the process at little or no added cost, yields of "higher" grades would improve.

Then the number of grades sold could be reduced, the customers advised that they were, in some cases, receiving a better product for the same price, and the amount of "low-grade" material to sell cut-rate might be reduced or even eliminated.

Using the same or similar equipment and tooling promotes the ease of working through problems associated with improving quality. It reduces the total maintenance effort and increases the ability to understand equipment and tooling in depth. One way to think of setup time reduction and setup consistency improvement is as standardization of the setup—a combination of equipment, tooling, method, and maintenance.

Standardization is improved by doing more tasks automatically than by people—provided the automation is well designed. Software glitches and nonstandardization are a well-known problem, and automation commonly stops as only islands of automation, which cannot communicate with each other unless considerable effort has been expended to make them compatible.

For operations involving many people, standardization forms a base for further improvement. Improvement ideas used by only a few people are not really improvements because they are inconsistently practiced. Furthermore, if the work practices of many people are inconsistent, no one knows where they stand in order to make improvement. (Failsafing is one way to standardize operating procedures.) A plant in which operators coming on one shift immediately adjust the setups of the outgoing shift has lack of standardization—no uniform concept of what the correct setup should be. The same is true for other kinds of work in which people do not agree on how the same tasks should be performed.

A good sign is prominently displayed operator instructions. The instructions they actually follow are the real standard. Standards books kept somewhere out of sight are apt to be out of date at best. If operators understand the correct method of work and are themselves responsible for seeing that the instructions visibly posted convey that to another operator, consistency of work methods is improved and the latest changes are incorporated into them. (Visitors in Japanese plants frequently see these instructions but miss their significance. They are only kept up if operators work together in maintaining them and if the supervisor tries out their validity by rotating operators from time to time.)

Keeping standard operator instructions current is a task easily underestimated. In many plants these are voluminous compilations, sometimes the largest mass of data the plant must deal with. Keeping them current is a quality problem in itself.

Attack the problem by reviewing the purpose of the instructions and the divisions of responsibility for them. Organized as a central review and data processing task, they can be overwhelming. Decentralize the responsibility. Divide the standards so that there are general plant instructions, departmental instructions, area instructions, and finally instructions specific to individual operations for specific products. The last category is the one that becomes voluminous. By making the important instructions visible, everyone begins to learn what is and is not important about operator instructions taken as a whole. As a result, the methods for keeping them current became simpler and standardized.

NOK, Inc., the company whose quality system is diagrammed in Figure 3–2, did not begin true statistical process control until recently (neither in Georgia nor in Japanese plants), yet it has shown steady progress in reducing defects. The company concentrated on seeking the causes of quality problems, taking corrective action to reduce variance, and standardizing results. The plan is to continue by adding statistical process control to the arsenal of methods for finding the causes of quality problems.

The relationship of standardization to quality is greatly underestimated. Standardizing a better practice, once attained, forms a new base for further progress. Without it, progress may easily become negative. With it, the first move in evaluating a quality problem does not have to be discovery of "what is really being done now."

CONCLUSION

This entire chapter is but an overview of the thinking necessary to attain superior quality in manufacturing. A poor quality situation cannot be turned around with one or two ideas, although a good start can be made in improving quality before developing expertise in statistics.

Attitude is most important. If a company is hell-bent on perfecting quality, it will persist in both discovering the methods to do so and in finding the means to employ them. Crafting a quality approach for a manufacturing company is not a short-term task. It requires making quality a centerpiece of the company culture. Many of the ideas are simple enough, but none are quick and dirty to put into widespread use, so quality must matter more than turning a quick profit.

However, quality is not only free, it pays dividends. As we shall see, many methods for improving quality also improve productivity. Total quality is the basis for developing just-in-time manufacturing adapted to the particular circumstances of a specific manufacturer. To do it on time, everytime, it must be done right.

CHAPTER FOUR

Just-in-Time Manufacturing

To properly understand JIT manufacturing, the objectives need to be reviewed from time to time: eliminate waste, improve quality, shorten lead times, reduce costs, improve morale, and continue improvement. The subject cannot be discussed without reference to techniques, but techniques are not sacrosanct. The importance lies in attaining the objectives in context of the necessary manufacturing tasks.

Begin with people. No matter how automated, the equipment improvement begins with people, and those who transform their companies regard the physical changes as manifestations of success with the workers and managers.

Just-in-time manufacturing may start in many ways, but the improvement of operations cannot long proceed without the development of skill in quality improvement as presented in Chapter 3. Otherwise, the technical obstacles block progress. Start with workplace organization.

WORKPLACE ORGANIZATION

A first impression of workplace organization is that it is shop-level piddling to which only workers and supervisors need pay attention. Staff and management should concentrate on systems and policy, the more complex and advanced ideas of just-in-time. Just-in-time actually begins with the basics, which are not just shop methods, but fundamentals of company policies and practices as reflected in the shop activity.

Workplace organization starts in the plant, but its reverberations extend to an entire manufacturing organization. The purpose is to clarify problems wherever they may exist. It is described in five steps here, but it is sometimes referred to as a four-step method with the last two steps combined. The five steps are: (1) clearing and simplifying, (2) locating, (3) cleaning, (4) discipline, and (5) participation. Much of the power is in the first two steps, and the last three closely support them.

1. Clearing and Simplifying. Remove everything not needed for production activity in the near future. Much of this may be work-in-process inventory, but *everything* includes much more: excess equipment, tooling, gauges, supplies, personal effects, and, of course, rubbish. While this should make some cosmetic improvement, the intent is not to create a showroom, but rather a clarified work environment, and it should go well beyond just fleecing up the trash and doing some painting.

The initial issues may be confined only to the shop floor, but stopping at that point misses the objective entirely. Press on until management issues arise—as many management issues as everyone can stand to address at one time—and deal with them. Typical of such issues are the following:

"Material on quality hold." This phrase refers to material kept on the floor pending a decision to scrap, rework, or use in substandard condition. The first issue is why making such decisions takes so long. The second is why the decisions had to be made. In short, the practice makes visible the quality issues that otherwise may be deferred.

One story (probably apocryphal) is that in a plant beginning workplace organization, the supervisors and staff had agreed in advance that quality hold material should be moved to the quality control area and placed in a prominent location, signaling that action should be taken. However, as the bins and boxes accumulated, they inundated the quality control area. The QC manager, seeing much more piling up than had been anticipated, ordered that it be moved back to the shop floor—a move that irritated the general foreman.

The plant manager stepped into this with dramatic flare. He took a box of "quality hold" parts to the parking lot and called a group of bickering staff and supervisors to an outdoor meeting. Following a speech on rededication to quality improvement, he poured gasoline on the suspect parts and set fire to them, saying, "You don't need to make a decision on these, but you ought to get on the rest of them right away!" Presumably, more than one fire was lit to get on with total quality.

Backup equipment. Question the reason for any extra equipment. More than likely, some is present not just to handle an overload but because the debugging of new equipment is dragging, or the root cause of chronic malfunction is not being addressed, and the onus for corrective action is in manufacturing engineering, maintenance, tooling, or elsewhere than with shop personnel.

Gauging or other instrumentation. Another frequent issue is the accuracy and calibration of measuring devices used in production areas. Some are not correct for the purpose. Some do not calibrate the same as master gauges used elsewhere, so they are not used. Parts are taken to the master gauge for correct checking. The issue comes to a head when it is suggested that useless measuring devices be removed. The desired outcome is an improved program to calibrate and maintain instrumentation on a regular basis.

Tooling. Workplace organization pertains to tooling, jigs, and other such devices, and so the activity carries into tooling areas and maintenance shops just as on the plant floor. Tooling to be used right away should be on hand and ready for use. Tooling that will not be used or that has indefinite use should be removed. One resulting project is to review the system for storing and classifying the tooling. What is it, where is it, and when will it be used again?

Schedule stability. Workers and supervisors asked to remove everything not needed in the next few weeks immediately have a legitimate question: How do we know what will be required over the next few weeks—for sure? This question brings up the issue of the appropriate planning cycle for a particular operation, which may or may not be reasonable for a given shop as it currently stands. The objective is not to cater to the desire for absolute certainty in the shop at the expense of serving the customer, but to review what must be done to serve the customers properly—all the customers.

In other words, the meaning of flexibility in meeting the market has to be converted into a specific set of instructions. This will vary from company to company, and from plant to plant within the same company. For instance, one company may be in a capital equipment business in which shop work is only a portion of total projects with customers, most of which have total lead times extending well over a year. Another must respond to any order with a custom-assembled product within days. A third makes to stock.

The needs are very different, but clearing and simplifying must lead to detailed specifics for improving service to the customer. Clearing and simplifying requires intensive effort to define the operations necessary to the customer—an interaction between company policy and the physical activity that carries out the policy. It is a company clean up as well as a shop clean up.

2. Locating. The principle is to have a place for everything and keep everything in its place and ready for use at any time. This suggests the mentality of a fire department or a fighter-interceptor base, and if one thinks of production as fighting a war, the analogies hold perfectly.

The general rules of location are common sense. Return things frequently used to a standard, fixed location, handy for use. Keep things used together grouped together. Devise general location rules for all operations, and specific ones for specific operations.

However, pursuing this rigorously implies more than such commonsense thinking. The detailed layout of a work area is the physical imprint of how work is actually performed, not only by a worker tending each operation directly, but by all the indirect and staff workers whose diligence, or lack of it, contributes to the tasks necessary for the worker to perform.

If the location of everything is done in detail and kept standard in practice, the worker must be involved. A staff group cannot possibly have time to do this for every operation, and even if they did, workers will modify the methods as they see best anyway.

If workers are to finish detailing workplace organization and work methods, a degree of worker involvement is necessary. If they are to do it well, considerable worker involvement is necessary, and they must be developed for it. Workers should become practiced in the basics of quality improvement and methods improvement, and they should become a team capable of standardizing on a particular method in practice while searching for better. If they can do this, workplace organization is detailed; if they cannot, workplace organization is coarse.

Workers detail it, clean it, and maintain it. It becomes a quality zone into which they should permit no defects to enter and from which no defects should knowingly pass. Responsibility for an area becomes responsibility to operate an area. Since more than one person typically works the same station, it is a team effort. (Automation does not change the principle of this.)

This innocent-sounding idea is revolutionary. The worker presumably wants to do a good job, and the role of the supervisor shifts to coordinator, facilitator, and *critic*. (Worker responsibility is not a religion of low expectations.) The role of staff shifts more toward coach and teacher, which is harder for them to accept than for workers.

The initial stages of workplace organization may proceed rapidly, but soon it begins to go more slowly for two reasons. First, workplace organization can go no faster than staff, supervisors, and workers can be developed for it, and people do not change their approach to work overnight. Second, problems off the shop floor must be addressed, and these problems take more time to solve than most shop problems do.

Furthermore, workers must communicate. If two workers operating the same machine want the detailed positioning of items in a different pattern, they may be operating the equipment differently—which in turn can be significant for both quality and maintenance. Workers without much responsibility for their area may not focus on the work itself if they make suggestions for improvement, but workplace organization is a catalyst for productive discussion. Ask workers to detail their work methods, step by step, and compare the results between those who work the same job. Such comparisons may turn up opportunities for improving both quality and productivity. Keep revealing the hidden problems.

A major objective of the location patterns is *visibility*—ability to see conditions at a glance at all times, with exceptions standing out because something is not where it should be. After nine months of workplace organization at Hewlett-Packard's Puerto Rico plant, the plant manager said he could tell if there was any kind of trouble just by stepping onto the plant floor and looking around a few minutes. More important, so could anyone else familiar with the place. The need for written reports or computer reports became minimal.

However, the principle of visibility goes deeper, applying to staff and management work as well. More than just shop rules emanate from clearing and location. Changes come in such things as drawing standards, how to state specifications, creating instruction sheets, methods of scheduling, and reporting. As the reform spreads, it eventually touches marketing, accounting, engineering, and all other functions of the company; but for this to happen, the proper organization climate must be created.

3. Cleaning. A clean workplace makes the unspoken statement that quality work is expected. A work area should be clean relative to the type of work performed. A foundry can hardly have the same standard of cleanliness as an environmentally conditioned room in which integrated circuits are manufactured, but both should strive for the same standard: Clean enough to avoid quality and maintenance problems. Clean enough to avoid health and safety problems. Clean enough to promote visibility as described earlier.

Cleaning is part of preventive maintenance and part of visibility. Those who always clean their own automobile notice flaws unseen by those who do not. Even if they do not repair defects themselves, they notice them. For the same reason, production operators should clean their own equipment, no matter how automatic, as much as possible. Operators should learn every quirk of their equipment as thoroughly as possible. An operator who cleans feels pride, and one knowledgeable about the equipment is less likely to damage it through carelessness or ignorance and likely to use it with more skill—more so for operators who accept that the equipment is *theirs, they* are responsible, and they have no backup. (The author has personally observed operators in settings opposite to that described deliberately jam equipment just to take a break while maintenance undid the damage.)

4. Discipline. The term connotes Marine boot camp, but production managers, workers, and staff are engaged in something more complex than close order drill. Yelling will not do. *Discipline* is consistently working by the rules and standards of clearing, locating, and cleaning. It comes from training, understanding, and what behaviorists much more politely term *reinforcement,* for everyone will be careless from time to time. Learning to practice workplace organization by habit does not happen overnight—but it should not take years, either.

Discipline is practice and correction, practice and correction. Think of it as coaching, not drill instruction. Everybody learns to live by both the general rules and the special area rules that should come out of workplace organization. It helps discipline if workers have the opportunity for input when rules are made and revised.

5. Participation. The changes on the shop floor should be basically understood by everyone, and everyone participates in them in some way. If workers are given responsibility to maintain workplace organization and a visibility system with it, "visitors" to the shop floor should not unwittingly disturb it. Visitors include repairmen, engineers, contract workers—even executives. Nothing wilts morale quicker than an executive violating shop rules, but executives can also wilt morale by not following through to correct the nonfloor problems that prevent workplace organization from materializing.

Participation in the broad sense is participation in all activities that promote the effort. A company should become a beehive of projects to improve production, design, paperwork, tools, equipment—even marketing.

Some of this was first explained to the author by Mr. Tanaka, executive vice-president of Tokai Rika, who explained that much of his company's benefit from just-in-time manufacturing was really derived from developing good workplace organization. The first time Tokai Rika tried its revolution it failed, and Mr. Tanaka accepted much of the responsibility for that failure. He and other top executives had not understood the scope and intensity of the approach. When they began again, executives went onto the floor of their largest plant and, in front of workers, began to personally sweep the floor. The workers got the message because they understood the intent, but someone unschooled in workplace organization would have thought they had come unhinged.

About six weeks after workplace organization began in one plant, a contract repair crew on the roof began to vibrate dirt into a production area. The workers were indignant. Six weeks earlier they would have walked off until management solved management's problem, but now this was *their* area, and they wanted action. The contract work stopped immediately, and the company paid extra for the contract crew to work overtime after the workers had covered up the production area at the end of the day. The result was worth every penny of overtime.

Morale is an intangible benefit of workplace organization. It pays a big return even if putting a number on it is only a mushy guess.

VISIBILITY

Visibility is unwritten, unspoken communication, not only of shop floor conditions among shop floor people but a road map of company conditions to all who read the physical signs. Visibility is a minimum-cost database made up of the real thing. Its major purpose is instant response from anyone who should take action. Visibility is akin to immediate feedback.

On a shop floor, visibility is attained by many means:

Posted schedules. A common form in a JIT plant is a board or electronic screen showing the current day's assembly schedule and completions to schedule. Everyone can see where they stand. In fabrication areas, schedules should be posted for all to see, not just printed out on a crumpled paper in a supervisor's pocket. That way everyone knows what the supervisor knows.

Signal lights. If a machine malfunctions, operators and maintenance hop to it. A counter used on a machine can trigger a signal for setup change, tool change, quality check, or whatever. Sensors can trigger lights indicating unusual changes. This idea is not new. Equip a shop floor so that it signals those responsible in the same way as lights on the control panel of a refinery.

Charts, logs, and goals. Posted charts emphasize what is important, and the charts should not become yellow. The purpose of the charts is to convey matters of current direction and action. Last year's quality award belongs near the entrance—an archival area, not an action area.

Layout. While dirt, noise, material handling, and many other factors are major factors in plant layout, visibility certainly belongs among them. Just designing an assembly line or fabrication line in a U-shape can have this effect because the back end of the line can see what is happening at the front. Also, locating feeding operations near the points they service has an effect. At one Hewlett-Packard plant, a printed circuitboard subassembly area was relocated within sight of the final assembly line it served. Defects went down and morale went up as workers could identify the final instruments into which their boards went.

Strict limits on inventory visible to those working with it. An operation running out of material sends up a flare in any plant, but having too much inventory may not cause alarm until well after it is acquired. Assuming a flow shop, designate a spot for each part number and set up a visible limit so that once production is stacked to the limit, it ceases. As material is withdrawn from the stack, one can see the "clock ticking" until production is next required.

Inventory should not be hidden. When operators finish a part, they should put it in a specified location for pick up and for everyone to see. Putting it on display has a powerful impact. Each operator is fully responsible for the quality of his operation.

Some companies overdo visibility at first. The shop floor looks like a disco parlor and has to be toned down. True visibility produces a response. It has a purpose: *Visibility creates effective, immediate feedback.*

This use of inventory as a stimulus to action is the most important aspect of just-in-time methods of production. The action is everywhere in the company, not all in the shop.

For example, at Hewlett-Packard's Greeley Division, modest progress had been made in revising the assembly operations with limited inventory when an incident stopped the line. A buyer and an engineer had not correctly communicated a specification to a supplier who failed to deliver. Everyone in the plant knew the line was down, which part caused it, who the supplier was, who the engineer was, and who the buyer was. Said the buyer, "I understand the reason for limiting inventory in the plant, but we really should carry a little extra for things like this. It's embarrassing."

The buyer and the engineer did not enjoy the spotlight, but management did not increase the inventory limit. Visibility on the shop floor is visibility into staff and management as well, and the time limits imposed by inventory limits should cause *everyone* to do their jobs right the first time.

Visibility extends to management practice away from the shop floor. For example, during a study of JIT practices two years ago, several teams of people spent two days on the shop floor of an electronics company studying how to improve work methods and material flow. The third day, a few from the group visited a supplier and were surprised to see an advanced automatic inserter scheduled for delivery to the same area that had been studied for the past two days. No supervisors, workers, or managers had said anything about it. Obviously, few people directly involved with the work knew of it. Such surprises, even if they are technically good, destroy the morale of workers and supervisors struggling to improve current equipment. A staff project did not have high visibility, so it was not factored into everyone's thinking.

Visibility is important even in office work. Communication in a project is one major problem alleviated by setting up a "war room" and work areas to display current status and problems at a glance. It cuts down on the memos and meetings.

Not long after Chrysler began to strictly ration the amount of inventory between plants, its Twinsburg, Ohio, stamping plant went on strike. The reason: The plant was overloaded. People were working 70-hour weeks, and no relief was in sight. They walked out.

A *Business Week* editorial called for the auto industry to rethink its then-limited embracement of just-in-time practices. After all, Chrysler assembly plants were shutting down, turning off the company's revenue spigot at a time when the financial community was still very nervous. However, Chrysler managers understood. They had to go to Twinsburg and work out the problem right away. It could not be ignored.

Many strikes are preceded by strike-hedging inventory, and the unions build up a war chest. The existence of both allows both sides to dance around for a long time until they must become serious. The existence of this inventory clouds the real problems.

Extending this a bit, if a manufacturing company is to thrive with very low inventory, then the timing of all work is critical, and little time can be wasted in protracted confrontation between various parties: management and labor, management and suppliers, or even management and management. In a sense, immediate survival issues tend to stand out above ones that in the end would have seemed petty.

PRODUCTION CONTROL SYSTEMS

The classic two-card production control system used for the *Kanban* method of control is diagrammed in Figure 4–1, and a one-card system is shown in Figure 4–2. Figure 4–3 diagrams a production network using a "pull system," but the method of sending pull signals is not specified. Material is pulled into a final assembly area from feeding operations as it is needed. In turn, each supplying operation pulls material into it as needed, and so on down through the series of operations until one reaches raw material. The system even reaches into the operations of suppliers if they have been developed for it.

Rules for the two-card system are simple but strict:

- One card, whether a move card or a production card, represents only one standard container for one type of part, and exactly the same number of that part goes into each.
- As soon as parts start to be taken from a standard container at its point of use, the move card is detached from the container. It is returned to the supplying operation as authorization to bring another standard container of the same part.
- Containers filled with parts should await pick up at clearly designated locations next to the supplying operations. When a full container is taken, the production card is detached and left at the supply point. The move card is attached to the full container and taken to the point of use.
- The production card left behind is authorization to the supply operation to make another standard container of the same part and to leave it with the production card attached at the same outbound location, ready for pick up.

In other words, the system of material control is to replace exactly what has been consumed and no more. The two-card system is actually one of the more complex ways of controlling this. Between two operations within sight of each other, no cards are

FIGURE 4–1 Two-Card Pull System between Two Work Centers

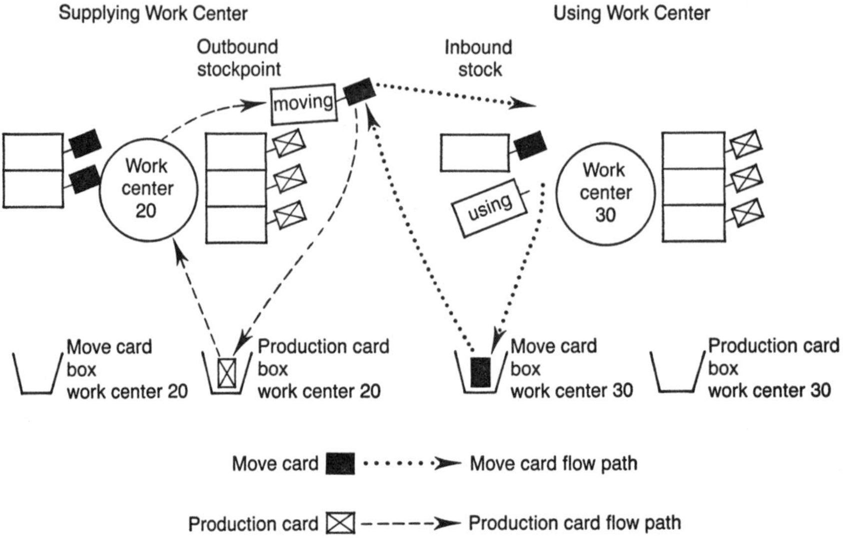

Cards go on standard containers. Assume one of these standard containers is always to be filled with exactly 10 parts.

There are three move cards:
One on the container waiting, one in the box, and one having just been placed on the container starting to move to work center 30.

There are four production cards for this same part number:
Three on containers waiting for pick up in work center 20 outbound stockpoint, and one having just been removed from the moving container and placed in the work center 20 production card box.

There are seven total cards.
Thus, if all rules are observed, a maximum of seven containers could be filled at any time. At 10 parts per standard container, a maximum of 70 parts could be in work-in-process inventory, plus whatever might be left in the container being used.

needed at all—only a strict restriction on the inventory between them. This can be done by marking a space between operations, sometimes called a *Kanban square*. If all squares are full, stop. If not, fill them up. Some squares hold only one workpiece. The same affect can be achieved by operations connected by a conveyor so that the inventory is limited by the space on the conveyor. However, conveyors themselves can hold too much and become repositories of excessive work-in-process inventory.

A one-card system uses only the move card. Space restrictions or visible limits sufficiently restrict the quantity of each part in its

FIGURE 4–2 One-Card Pull System between Two Work Centers

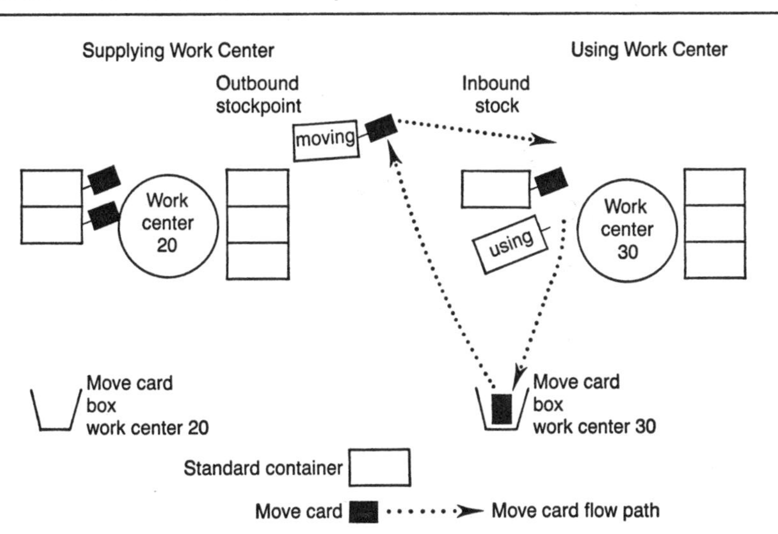

One move card goes on only one standard container. Assume each standard container will be filled with exactly 10 parts. One card authorizes movement of one standard container filled with only 10 parts from work center 20 to work center 30.

There are three move cards:
One is on the container waiting to be used. Another has just been removed from the container starting to be used and placed in the box. The third has been taken from the box to work center 20, where it is attached to the container starting to move.

The visibility (or Kanban) rule for the same parts at work center 20 is to place them in standard containers stacked no more than four high. (No production card is used. It really is not necessary unless the delivery end of the production process must in some way provide a signal to the starting end.)

Four standard containers at work center 20 plus three move cards adds to a maximum of 70 parts that could be in work-in-process inventory.

If the containers themselves circulate between work centers 30 and 20, the containers appropriately marked might themselves serve as the pull signals. If no more than seven containers were used, the maximum work-in-process would still be 70 parts.

location specified for pick up. It the move cards are in effect permanently attached to the containers, the returning empty containers serve as signals to be filled up.

Replenishment signals sent electronically to supplying operations eliminate the time required to return the cards. This procedure is acceptable if an upper limit on inventory can be maintained.

FIGURE 4–3 Materials System Overview of Pull System

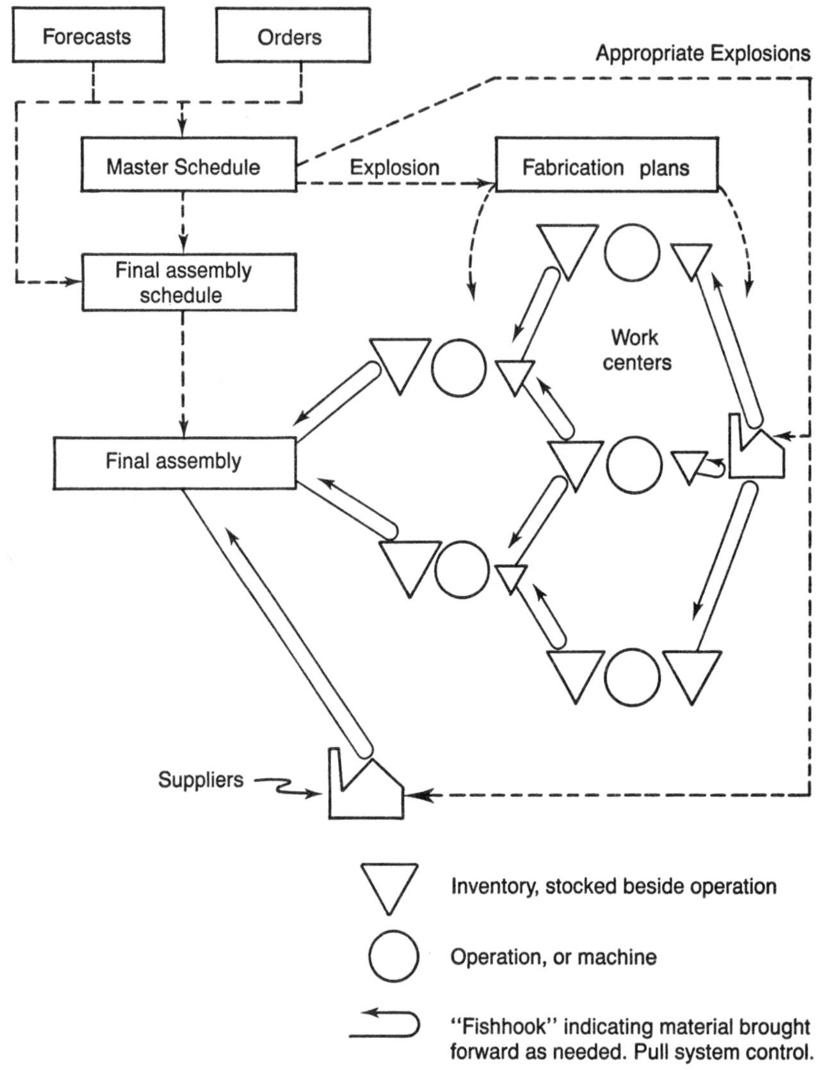

Forecasts
Orders
Appropriate Explosions
Master Schedule
Explosion
Fabrication plans
Final assembly schedule
Work centers
Final assembly
Suppliers

Inventory, stocked beside operation

Operation, or machine

"Fishhook" indicating material brought forward as needed. Pull system control.

Cards and containers can be equipped with bar codes or even with smart chips, which can be electronically coded with part number, where from, where going, and other information. These possibilities are so interesting that they may distract from the primary goals of manufacturing excellence, which do not include running a complex production control system. Unnecessary complexity is waste.

Pull Systems Are Really Just One Major Manifestation of the Principle of Visibility. Promote direct, immediate visibility of problems and conditions to the direct, hands-on people who can take immediate action. (An aged inventory list from a computer system is not immediate feedback.) The most important rules of pull systems promote this.

1. Provide simple, visible, enforceable means to limit the inventory of each item. Limit the number of cards that circulate or the number of containers that circulate. Restrict the space available for inventory. Mark it clearly.

2. Make only what is necessary. Observe the inventory limit for each item. Depending on the specific arrangement, make nothing without authorization of a card or container or do not overrun the space provided for inventory.

If an operator sees that all space is filled or that there are no cards to make any other production, *stop*. Do something else useful, but make no production before its time.

This violates the instincts of both workers and managers. The siren song is to keep production going on the theory that it will become useful sometime. However, the principle of visibility is just the reverse. If production is unnecessary and an operation stands idle, this overcapacity is immediately obvious, whether temporary or not. The idea is to develop the operation and to run it to do only what needs to be done.

3. Locate each part in only one place. Put material only where it is supposed to be. The same part may be used at many operations in a company. It may also be produced by several different operations, so this rule pertains to each operation producing that item.

If an item is finished, put it in the spot designated for pick up. If something is wrong with it, start corrective action immediately. If you do not know how, get help. In any event, do not hide the problem by putting material aside until something can be figured out.

The quality incentive is powerful. In fact, the next operation is the worker's customer and should be regarded that way. The objective is to always be figuring out how to serve the customer better.

4. Maintain the limit on inventory for each item at a level that stimulates improved performance. If an operation seems capable of improved performance (less waste), reduce the limit. If it runs into trouble (hopefully temporary), increase the limit in a way that calls attention to the problem. For planning purposes, formulas and mini-simulations assist in estimating the required inventory levels, but at bottom this is done by judging what an operation can presently do.

If an operation seems to be running smoothly, inventory can even be decreased slightly to point out problems that might otherwise remain unseen. Again, improvement counts.

Not managing inventory limits is one of the most basic and prevalent errors made in so-called just-in-time production. Mere use of a pull system is not enough. That is no more than a reorder point inventory system. This error is easy to spot on a shop floor because the workplace organization gives it away—inventory locations and limits are vague and sloppy instead of sharp and defined.

Managers should recognize that limiting inventory restricts the amount of time people have to respond to problems. Thus inventory levels are a measure of people and process development. If managers and workers have not developed so that they know how to respond to this time pressure, the result is frustration, just as if they were pressured for performance by any other means, but did not know how to respond. More than one person subjected to this pressure too quickly has had to have personal counseling to alleviate anxieties. At bottom, *limiting inventory is a people system, and it is the development of people that counts,* no matter how automated the process.

Why a pull system? Wouldn't another work as well for the same purposes? The pull system is simple; a computer may help but is not necessary—for the shop floor part of the system anyway. Most important, however, the outbound stock or simple presence of the intervening inventory between two operations is itself a signaling device. If parts are made and moved rather than made and held, the feedback signaling to the producing operation is not so simple. The producer somehow has to know to have material ready when the using operation wants it. Using the outbound stock itself as the signal is simple.

More important, users of material are not overrun with what they do not need, and so they do not have to store the excess. By keeping most material in outbound stockpoints, users may send for a small amount of what they want when they want. Suppliers instantly see what was taken—immediate feedback of real demand—eliminating as much advance forecasting and delayed feedback as possible.

A properly developed pull system relates inventory to time through visibility. Inventory becomes a clock, as illustrated in Figure 4–4.

For operations purposes, think of inventory as time whether it is repetitive and convertible to a pull system or not. Thinking of inventory as time stimulates thinking about the operations—and waste—filling that time.

The most basic requirement for any pull system to work is that each operation know where to send for material. Therefore, the sequences (or routings) of material must be clear and constant. Knowing what the next job received at a work station will be is not always necessary as long as workers know where to get it. If material does not flow in a standard path through production, a pull system will not work. Standardizing flow to make the system work makes changes in production that eliminate waste in themselves.

Another requirement for a pull system to work well is a reasonably level workload and a repetitive flow of material. This schedule must be created. More than just a numerical exercise in scheduling, the synchronized schedule depends on development of production processes for it, on the linkage with marketing and distribution, and on considerations in product design.

A pull system will not work if there is no repetitive potential in a production process. That condition occurs if products are unique and custom engineered, or if demand is lumpy and irregular, as is often the case for low-volume capital goods products. A production process on the edge of its technical capabilities is also difficult. It needs basic development before refinement—although this situation can be used as an excuse by those who do not wish to try.

However, great progress can be made in production settings that do not lend themselves to synchronization. The basic thinking is the same, but visibility methods cannot include a pull system.

FIGURE 4–4 Inventory as a Clock

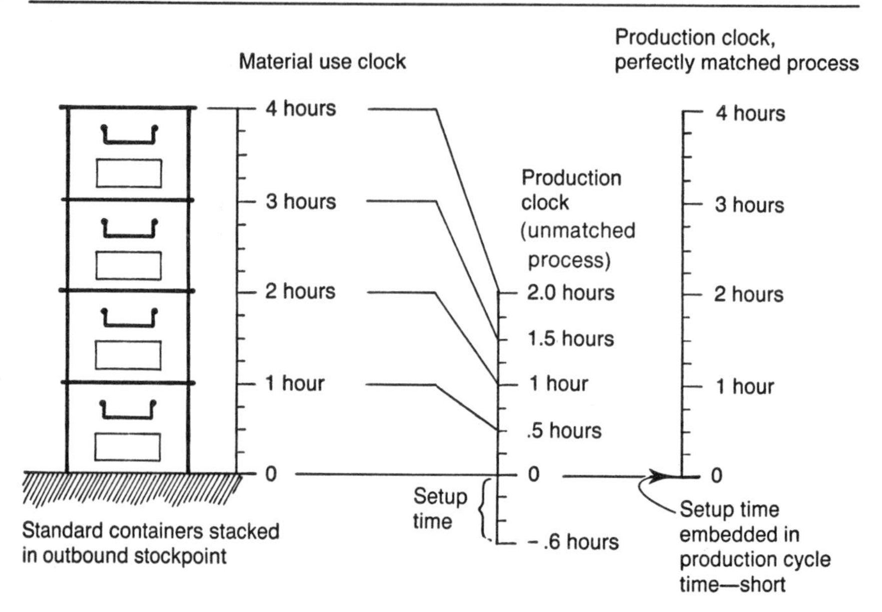

If material is used at steady rate and one hour's use is in each container, the stack is a clock.

In this case, each container represents half an hour's production, once set up.

Imbalance between use time and total production time, including setup, is filled with inventory. Same machine may make multiple parts.

In this case, no reason except transport timing exists for having inventory at all.

A perfectly matched machine may run slower.

Relate Inventory to Time

(Inventory turns is a financial, not operational, measure.)

Seconds on hand:	Automatic transfer.
Hours on hand:	Synchronized operations.
Days on hand:	Most repetitive manufacturers can start here.
Weeks on hand:	Most job shops can start here.
Months on hand:	Very slow cycle production. Few should use.
Years on hand:	Cemetery.

TIMING

The time between recurring activities is commonly called *cycle time,* a phrase used throughout industry in different contexts, depending on what is being timed.

- In materials work, cycle time frequently refers to the time between orders or to the time material is in house—dock to dock. In fact, materials managers are likely to think of JIT in a broad sense as reduction of cycle times or management of cycle times.

- Cycle times also refer to planning cycles. Business plans are often revised monthly. Accounting statements are prepared quarterly, so a quarter is an accounting cycle and a fiscal year is a longer accounting cycle.
- Industrial engineers use the term *cycle time* to refer to machine cycle times or to the times between completion of units on an assembly line. Sometimes it refers to the work cycle time of a person doing a repetitive task. (Cycle time is the inverse of production rate.)

Regular timing benefits both quality and productivity. As we observed in the case of curing rubber, a standard delay between mixing rubber and curing it provides consistency in the final product. This same phenomenon is important whenever time delay is a factor in consistency, as with drying paint, allowing oxide to form on aluminum, preventing oxide from forming on steel, and so forth.

Furthermore, if an end-product production schedule can be constructed (and actually held in practice) such that as many materials as possible are consumed at a steady rate, the materials can also be fabricated and delivered at a steady rate. This is the basis for classic JIT production and delivery. If material is not used at a steady rate, a complex flow of material becomes confusing, and the information load trying to get the right part to the right place at the right time becomes very difficult to manage. Computers may help manage irregular timing, but understanding what to do takes extra effort, and extra work is waste, whether computerized or not.

A steady, level pattern of material use also forms the basis for doing many production operations in a steady, repetitive pattern. Many activities are directly or indirectly related to moving, converting, or measuring material. The following are some of them:

Material handling.
Material transport (between plants).
Preventive maintenance.
Quality checks and calibrations.
Tooling.
Personnel schedules.
Even engineering changes and production trials.

Does an approach to problem solving appear because a company cuts inventory? Or does the resolve to be open in addressing problems allow the reduction of inventory? Attention to problem solving and people issues starts first. Otherwise, everyone is only frustrated when inventory is limited because they are not practiced as a team in methods of making and standardizing improvement.

One of the popular misconceptions of JIT among workers is that its objective is to speed up the work pace—put everything in fast forward. That is no more true than the belief that its sole objective is to cut inventory. In fact, too fast a work pace is not good because workers make too many mistakes, and a slow one is not desirable for the same reason. A good work pace sustains concentration but is varied enough in content to stimulate thinking while working. Monotonous work cannot always be avoided, but it is not desired by most humans.

However, the relationship of good timing to excellence in operations is not always intuitively understood. It has several aspects, which are described as follows.

Cycle Time Concept

Doing tasks at regular time intervals improves the possibilities for refining and standardizing them—up to the point where monotony sets in, and then perhaps automation or computers can be employed. For instance, setting up the same machine with the same tooling at about the same time every day stimulates people to make setup into a standardized, well-practiced activity, which aids quality.

Any work that can be done in a repeating cycle can be more effectively studied for improvement. Study by comparing operations from one cycle to the next, which can be done by basic or advanced methods. A basic approach is to time 10 cycles of an operation, such as manually loading components in a printed circuitboard. Both observer and operator should carefully note methods of work during each cycle, then review results:

1. What happened during the longest cycle time that could be eliminated?
2. What happened during the shortest cycle time that might be incorporated into each cycle?

Improving methods in this way shortens average cycle times and reduces the variance in cycle times. Doing either can improve quality as well as productivity. A name for this is *cycle time analysis*.

Any activity that can be cycle timed can be studied for improvement by cycle time analysis:

Single-station operations.
Man-machine interactions in cells.
Throughput times of material through plant.
Preventive maintenance procedures.
Material handling cycles.

Developing operations to produce at a steady rate creates the basis for repeating cycles of work in many activities. A single-station operation is usually simple. A man-machine cell analysis is a little more complex, but try to keep it simple anyway. Cycle time analysis of material throughput time is really analysis of many cycle times of all operations through which material travels on its way through a plant. Cyclic work and cycle time analyses apply to nonplant activities too, such as preparation to execute engineering changes.

Why engineering changes? If materials consumption runs at steady rates for one to four weeks, do not clutter smooth operations with disruptive changes. Hold them until it becomes time to change the scheduled pattern to better fit the market—a regular schedule change cycle. It is not so hard if almost everything goes in cycles and if inventory is reduced so that using up obsolete parts is no longer a major factor in engineering change execution. A plant in continuous improvement precipitates many engineering changes, so a company should learn to execute them well.

Getting many people to learn to use cycle times for synchronization and improvement is an important early step in mastering JIT in a repetitive production environment. Then, a simple final assembly schedule providing for steady parts consumption also provides a simple basis for many people to plan their work in cycles so everything is ready to deliver material at a steady rate. Material is delivered at regular times, tooling is refurbished at regular times, and everyone prepares to revise the pattern of work at a regular time. And, all activity done in cycles can be refined in methodology.

This whole set of ideas has been called the "cycle time concept," mostly by Japanese so far. According to this concept, a simple final assembly production schedule allows many supervisors in supporting activities to plan their work in sensible cycles that fit together. Cyclic activities are also easy to analyze for improvement. It is simple, and simple things work.

Refined further, schedules allowing repetitive, low-variance operation also provide the basis for simple, not complex, automation in indirect as well as direct production areas. Simple automation costs less.

Close Timing, Short Cycles. Lead times are usually thought of as lead times for material to go through a plant, and reducing materials cycle times has the easily seen benefit of reducing inventory investment. Other benefits are not so obvious, however. First is the elimination of waste. Materials cycle times are cut by improving quality, eliminating inspections, eliminating double handling, reducing setup times, and so forth. Every one of Shingo's seven wastes adds time to the processing of the product, much of it while material is just sitting. Inventory sitting is idle money; but worse, it incurs unnecessary cost.

One way to attack this is by flow charting a typical part through a plant, as in Figure 4–5. If this is done in detail, many steps are seen to be unnecessary. Eliminate the unnecessary. Briggs & Stratton has refined this approach so that flow charts are constantly being prepared to open the minds of staff and supervisors to new opportunities to eliminate the unnecessary. This is a process of people developing skill and knowledge about a product and the way it is made.

Slipping back is easy once a manufacturer has developed, so a limit on inventories provides an indicator of status and stimulates alertness to maintain progress. Reports of various kinds, including cost reports, can be fuzzed and hedged, but the existence of inventory does not lie. It is evidence of manufacturing proficiency or lack of it.

Tight visible control of inventory stimulates people once they have developed; but, for this to work, management at all levels must pay attention and read the signs for what they are. People change, supervisors rotate, and product designs evolve—not always in positive ways. A manager checking inventory levels and other visibility indicators on a daily round is not just checking the production workers to exchange pleasantries. He or she is checking the condition of the company as well.

As development proceeds, another objective of short cycles is flexibility—ability to respond to change. Ability to respond to changing customer demand sounds the opposite of running at a steady rate for several weeks, but it is not. If schedules change often enough, the next cycle's schedule can be very different from the current one, both in overall volume and in the mix of items produced. Inventory levels must reflect mix changes within a current schedule period. However, in most cases, companies are able to shorten the lead times given to almost all customers.

Reduce the total pass through time of the material train from months to weeks to days, beginning inside the plant, then working with cycle times through suppliers, which is more difficult. This is accomplished by eliminate the wasteful activities, improve the quality in operations, and synchronize the timing as much as possible. No company has complete control of all factors affecting the lead times of all material through the entire production network. Some material sources cannot be improved in the short run.

There are no miraculous transformations, so push the inventory levels as far back in the total process as possible, remembering that inventory existence, not ownership, counts. Once inventory starts to be converted into product, keeping lead times short as possible permits making what the market actually wants as nearly as possible. If the inventory can be pushed all the way back to "iron ore in the ground," it can be converted into anything—provided the process to do so exists.

TIMING AND FLEXIBILITY

Long lead times complicate an entire business, not just production operations. For instance, commercial aircraft production has a very long lead time, so aircraft orders start with preliminary feelers, then order definitions, then options to order, followed by purchase contracts hedged with cancelation clauses, review steps, acceptance procedures, and a tangle of financial arrangements. Few manufacturing businesses are this complex, but long lead times complicate life wherever they exist. After business practices become established, it sometimes is not obvious that a thicket of

FIGURE 4–5 Tracking a Part (trimpot, electronic)

Time	Distance	Step
		Arrive on truck.
		Wait to unload.
		Move to dock.
	30 ft.	Unload.
		Wait on dock.
First in-plant track point		Add to receiving report.
.5 day		Check against P.O. and bill of lading.
.5 day		Hold on dock.
	150 ft.	Move to inspection.
2 days		In queue in inspection.
	15 ft.	Take sample, move to inspection station.
		Inspect.
		Enter in system as good parts.
1 day	15 ft.	Move sample back to main lot.
		Delay before move.
	250 ft.	Move to stores.
.5 day		Delay before put away.
		Entry transaction.
	100 ft.	Move to bin.
4.5 days	560 ft.	In bin storage.

complexity grew up around some rotten old stumps of quality problems and long lead times—really waste.

The total manufacturing process should be developed for the necessary flexibility as well as for reducing cycle times of material. Quality, short lead times and flexibility uncomplicate things. However, the *total* manufacturing process must be developed for flexibility—understanding customer needs, product design, process development. Flexibility is important in several different ways.

Product Mix Flexibility. Ability to switch *quickly* from one product to another, one model to another, or one part to another

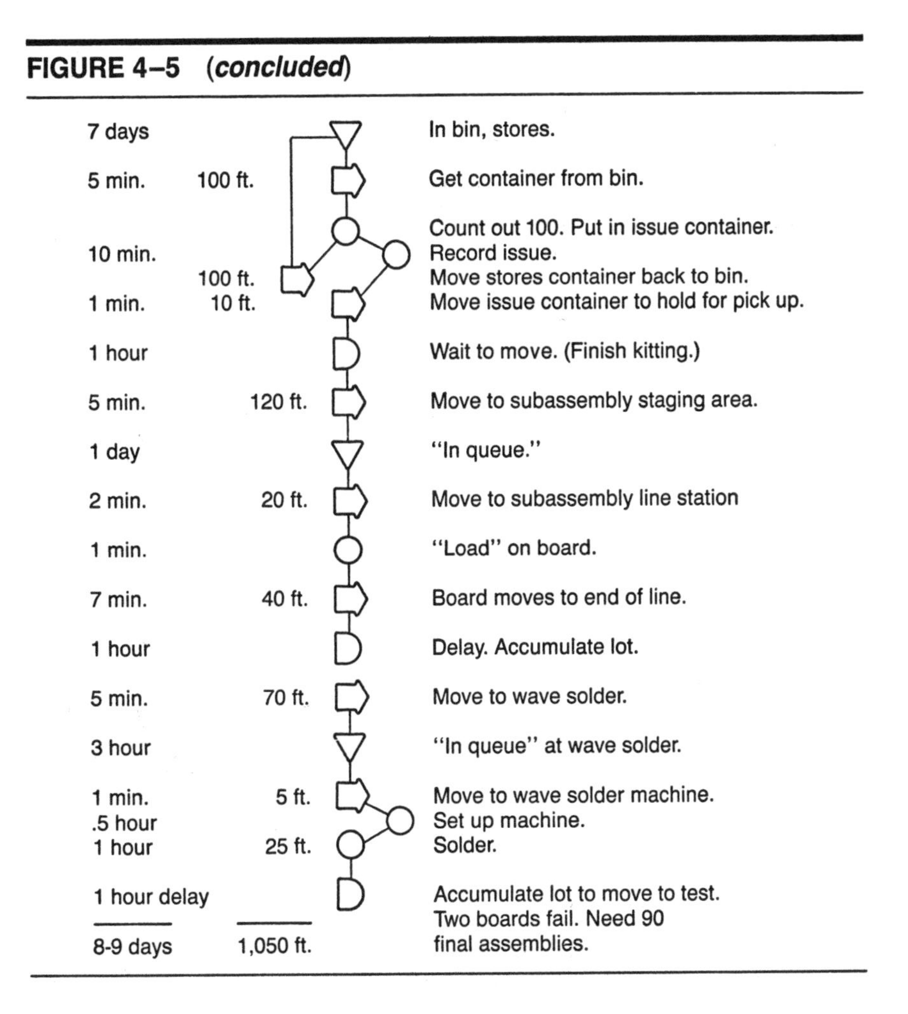

is essential for this. Low setup times allow this to be done at reasonable cost and also improve quality in the process. To attain flexibility, reduce setup times on all the machines through which material must flow. Reducing them on only a few machines in a long series does little good. Lot sizes remain large, lead times long, and eventually the incentive for keeping the setup times short fades for the few machines developed for it.

If lead times are short, all equipment must run at about the overall rate demanded by the market and must make the mix demanded. Therefore, smart automation is not high-speed equipment but rather equipment that will do what is needed at the time needed.

Standardize and automate the setup operations, not just the running operations.

Total Volume Flexibility. A plant capable of operating successfully over a broad volume range needs a low break-even point. Toyota is said to have a break-even objective of 30 percent of rated capacity. Break-even depends on revenue as well as cost, but one clear goal is not to invest much. Another is to create operations in which people may be added and subtracted, machines added and removed, and machines function over a range of speeds. All plants can exercise at least one option: increasing or decreasing the amount of time for production. Having numerous flexibility options involves people issues as well as hardware issues.

The situation to avoid if at all possible is an operation with only two speeds, on and off. When on, it must be kept running for a long time so that heats or other conditions stabilize to give good quality. That leads to decisions to run the operation just to be running it—which builds inventory.

People Flexibility. Inflexible workers have the same effect. Specialists stand around waiting for someone else to finish their part of a job. The same is true of specialists in staff areas, except people are sitting instead of standing. Specialists are necessary, but a company is better off if everything does not have to wait on them or keep them busy. Cultivate specialists who believe that less specialized work is not beneath them.

People willing and able to do whatever is within their capabilities when it is needed are the people side of JIT. They make balancing tough sequences of operations possible. They respond to volume and mix changes. They do what humans are by nature equipped to do—a variety of things. If production at one operation must stop, they move somewhere else. If the entire work force performs to only narrow job descriptions, one person transferring can domino a dozen others into unfamiliar jobs. A layoff of any magnitude can cause almost all workers to start a new learning experience, to the detriment of both quality and productivity.

Inflexible work rules are so detrimental to shop productivity that they have become management targets in union negotiations. However, management and staff inflexibility also takes a toll—perhaps greater but harder to see. Specialists are likely to have a narrow perspective of the relationship of their own job to the whole business, which complicates communication.

"Communicating" in various ways is a big part of business overhead. Much of the communicating is relating to others and to the *total* work to be accomplished. Job rotation and visibility are just as important in staff and management work as on the shop floor, and they are vital in adjusting to variations in volume and mix.

Engineering Change Flexibility. Nothing stays constant in product lines with short life cycles, and changes occur even on long-life products. The desired improvements in the manufacturing process seldom can be realized without many design changes. Design changes are a fact of life, and how they are managed is very important.

Collect engineering changes for execution at regular times. This becomes feasible with low inventory once the cycle time concept is in place, and it simplifies the communication about when to execute and whether a change was actually executed. (Most manufacturers are not exceedingly proud of their communications on engineering changes.)

Move each new design as far down the learning curve as possible before production starts; then press the revisions necessary for producibility in one or two short bursts as soon as production has begun.

All Japanese motorcycle companies introduce several new models per month into plants already producing a large mix. Each day after production is complete, some of the time remaining is used for trying the new tooling and arrangements for upcoming models. When it appears that production can go without a hitch, the new model is inserted into the regular production line up for the succeeding schedule month. Any necessary finishing changes are incorporated at the end of the first or second month of production.

Some kinds of production are much more subject to engineering change than others, for example, wiring harnesses. Design changes of many different kinds call for a change in the wiring harness. Therefore, that business must not only be just-in-time with production but just-in-time with changes in design. The ability to rework the fixtures used for harness layup must also be "quick change."

New Product Introduction. Development of skill in total quality and the many facets of JIT manufacturing may confer a competitive advantage—ability to work down the learning curve of a new product start-up very quickly. Timing here is of as much importance as anywhere.

Japanese companies measure progress by counting the total improvement suggestions made and implemented. Suggestions are a morale booster for employees but also indicative of much more, such as extensive practice of cycle time analysis. Employees of all stations who make and implement suggestions keep themselves sharp. If products and processes have matured, many suggestions may be rather trivial, but when a new product idea comes along, they are ready to pounce on it. Suggestions then are significant. (The current major concern of Japanese managers is how to improve the introduction of new products and technology.)

If a new product incorporates leading edge technology, it may be hard to refine for production considerations. However, many new products result from a series of small technological steps, not a great technology leap. Much design engineering ought be a highly integrative organizational activity to achieve fast, smooth introductions.

CONCLUSION

Just-in-time manufacturing is the pursuit of the very best management approach possible for a company to meet objectives from eliminating waste right on through continuous improvement. Usually it is seen as only picking up techniques or, at best, making an addition to the management methods known to the company. It really is a way to run a company.

Not all companies can use all techniques. No one should expect them to. However, most of the ideas can be used in some combination by any manufacturer dedicated to manufacturing excellence.

APPENDIX: WORKPLACE ORGANIZATION COMES TO INDUCTOHEAT: JIT IN THE JOB SHOP

The assumption is too frequently made that a nonrepetitive manufacturer cannot do much with just-in-time manufacturing. That

Target, periodical news source of the Association for Manufacturing Excellence, vol. 2, no. 3, Fall 1986.

assumption is being tested by Inductoheat, a small manufacturer tucked away in Madison Heights, a northern suburb of Detroit. True, Inductoheat is not into mixed model assembly or Kanban-style production control, but it is well along in transforming itself by much of the manufacturing philosophy promoted by the Association of Manufacturing Excellence.

Inductoheat is the quintessential job shop. They design, build, and install induction and resistance heating equipment, controllers, and power supplies. Well over 90 percent of sales is a brisk business in electric heating systems for heat treating and forging. Most jobs require some degree of custom engineering, and some are genuine technical development projects with the customer. Although the company is striving to increase the percentage of near-standard units in the product line, none are totally off-the-shelf, and it builds to order.

During the past five years, Inductoheat has revolutionized its total approach to operations, expanded its market to global scope, and changed its name. Until recently, the company was known as IPE Cheston. The product line is attractive for three reasons:

1. Induction and resistance heating makes it possible to generate multiple heat treat histories in different parts of the same workpiece.
2. These systems are the major means to heat treat in line with other operations on the same workpiece, thus eliminating a trip to central heat treat, which is of considerable interest to companies reconfiguring metalworking operations into cells.
3. They are pretty good at it.

John Stoll, vice president of operations (and an AME member), was reluctantly persuaded to part with the figures in Table 1. Privately owned, Inductoheat's owners brag no details in an annual report and regard all dollar figures as personal among their employees, their auditors, and themselves. Inside the company, cost particulars have become relatively open compared to five years ago. The owners wholeheartedly encourage John's shifting of responsibility to the hands-on workers, and they are downright pleased with the cash flow implications of the results so far. John just says that "things are going well," and he is spending more of this time nurturing newly acquired ventures in other parts of the world.

Table 1 compares 1981 with 1985. In 1981, the company began running an MRP package, which was part of its transformation,

TABLE 1 The Inductoheat Take-Off

	1981	*1985*	*Notes*
Sales index	100.0%	200.0%	Some of this was export!
Return on owner's equity	13.3%	24.2%	It was -2% in 1982.
Total inventory index	100.0%	50.0%	An increase of about four times in turns.
Stores inventory index	100.0%	33.0%	
Total employees	150	105	Sales/employee increased about 2.8 times.
Shop headcount	58	33	Five full-time supervisors now cut to two part-time.
Total space	94,000 sq. ft.	44,000 sq. ft.	Closed one of two buildings and leased it.

but results that year were average for the history of Inductoheat up to that time. In 1982, the firm lost money because of the economic debacles among its automotive customers, and in that year it began the changes in workplace organization, which stimulated other activities normally associated with the JIT philosophy. 1985 shows results after the operating revolution had taken off. (In fairness, it was also a better capital spending year in Detroit, but the company might not have been in so good a position to take advantage of it without operating changes.)

Inductoheat designs and fabricates in its own shop the items critical to its systems—the induction heating mechanisms and the electronic controller boards. Everything else is purchased for assembly. The materials system has approximately 12,000 active standard part numbers, and other items unique to each job are purchased. The shop consists of four departments in a single-floor plant:

1. Electrical.
2. Inductor.
3. Mechanical assembly.
4. Development.

Sales and engineering offices are in the same building, so all operations are now contained on site.

The story starts in 1980, when several managers attended an Oliver Wight MRP II seminar; as a result, they began shopping for an MRP package that was both simple and low cost. They found one for $15,000 plus $500 for installation, not counting their time, which was about four months of do-it-yourself. Most of the effort was in creating the bills of material and formalizing the inventory. The system began running in mid-1981.

It is a simple system with no extras. Bills of material rarely go more than three levels deep, so it costs little to run and maintain. It mostly explodes and backschedules and pumps out minimum paper: weekly time buckets, weekly regeneration, 26-week horizon, and the master schedule is end item job numbers at job staging time. A bill of material is created for each job, which is simplified by using modular bills for standard subassemblies and modifying them as needed. Jobs early on the horizon often have an incomplete bill, which is necessary for the long lead time materials, and the bills of material grow to completion as design evolves during the lead time. Everything is done with only one bill of material.

This basic system helped Inductoheat by applying computer system to the materials confusion factor, but operations were still amiss. The firm thought about supercharging the system horsepower with a shop floor control module and capacity planning module but decided on a different route. To see why, first look at the content of total customer lead time for typical Inductoheat orders:

Activity	*Estimated Range of Lead Times*	*Lead Time in Weeks*
General discussions with customer on problem areas	0–2 years	0–104
Concept and proposal	1–12 months	4–52
Proposals reviewed, reworked, and finally accepted	0–12 months	0–52
Detailed design	1–3 months	4–13
Fabrication and supplier lead times (a few items take longer, sometimes much longer)	6–8 weeks	6–8
Staging for assembly	2 weeks	2–2
Assembly	1–3 weeks	1–3
Test	1–4 weeks	1–4
Shipping, installation, and trial	2–8 weeks	2–8
Total weeks		20–246

Shop work is a small portion of very long lead time projects, whose duration, urgency, and technical innovation exhibit great variance. Much of the effort is determining how to meet the customer's specific need and translating that need at every stage of each job into hardware and methodology for use as a heating system. A due date is more than a target for suppliers, fabrication, and assembly. Tardiness or indecision in design can easily lengthen actual lead times for shop work, at the same time delaying the beginning of it until all must be done under time pressure. Worse, it can befuddle drawings, specifications, and shop instructions, and it does not help system installation and training for the customer either.

The condition of a shop floor reflects this. From it one can read the condition of management in places far from production. Workplace organization in the complete sense is reorganization of all operations because all funnel into the shop floor in some way. Inductoheat began shop floor workplace organization in 1982, following the five-step plan:

1. Remove everything unnecessary, not just inventory.
2. Create a location for everything.
3. Cleanliness.
4. Discipline.
5. Participation.

Personnel practices, supervisory methods, work methods, and many other changes accompanied this, perhaps as a direct outgrowth of the shop workplace organization, perhaps not. It is hard to distinguish the causes of change after the fact.

The company began by clearing out all excess material, equipment, tooling, and trash—five semi-trailer dumpsters full. To create a location for everything, the policy toward vendor deliveries was tightened. All material specific to a job was to be delivered within the two weeks established as the staging time for a job. At the start of the two weeks, a new job was assigned a cleared location, roped off in assembly, and a sign was posted designating the job. Standard parts were pulled from the stockroom and moved to staging. The delivery times were enforced by either shipping back material arriving early or refusing to pay for it early. All regular suppliers now understand Inductoheat means business with this rule, and, at its size, few suppliers would regard it as a gorilla exercising leverage. As a result of eliminating double-handling of material, total labor decreased about 30 percent for an equivalent job.

Previously, if some detail of design dragged or if a supplier were late, material was pulled for a job; however, the staging stretched out so some parts had to return to stores or be moved to make way for work that really was ready. The practice had led to inventory all over two different buildings on two sides of the street. After about a year of this, Inductoheat progressed down to where all operations were done in one building; soon after, only one building was needed for everything. In the meantime, the number of total jobs being finished increased.

The workplace organization, coupled with other actions, led to improved visibility in the shop. (*Visibility* is the ability to see shop status and problems at a glance, and shop visibility makes transparent many problems originating elsewhere. In a sense, major improvement in shop visibility opens the eyes to waste in many nonshop activities. The capital savings from space reduction occur but once; the benefits of improved visibility go on as long as the organizational discipline for it remains.)

A job schedule board is prominently posted in the Inductoheat shop. All fabrication is done for one of those job numbers. By MRP, material is acquired an appropriate lead time before assembly staging, and all material is issued to an assembly job number. No person or report must advise workers of priority of fabrication work. The job schedule board tells them. As soon as a new job staging area is roped off, no one must tell a worker to expedite a lagging part. The situation is obvious—in fact, so obvious that people on the floor are upset if nonshop activity is not going well. John Stoll and other managers read the same signs by walking through the shop, and they become upset too; thus the shop condition has become a communications medium of both management and labor.

At Inductoheat, engineering design details do not always work out as fabrication and assembly proceed. Fabricators cannot work metal as intended, and assemblers must modify parts so that the as-built configuration works. Simple changes they make themselves. Those critical to operation require engineering recalculation. To get through this, a simple rule prevails at Inductoheat: The design engineers must work directly with the shop workers fabricating and assembling their design—no intermediaries. That ended a time-consuming responsibility of management, and since there is well-posted schedule to meet, engineers and shop workers both learn a lot, and Inductoheat's total operations are much improved. Assembly times are estimated to have been whacked by 50 percent.

All workers are multiskilled. Inductoheat now trains them that way. Five years ago, specialists stood around waiting on each other to do their bit of the total project. Now the same worker does machining, welding, brazing, assembly, and even loading of printed circuitboards on occasion. One of the two supervisors loads PC boards during otherwise idle time. Loading boards had previously been considered one of the lowest status jobs, but no more. The key was to abolish *both formal and informal job status symbols*. Everybody comes dressed for work and pitches in doing whatever is necessary.

Inductoheat begins training new workers as machinists. Then they learn torchwork—cutting, brazing, and welding. From there they go to assembly, so all assemblers can make or modify basic parts as necessary.

Naturally, some workers have specialized tasks, and some have more skill than others. Three of the more skilled individuals are lead workers with engineers on the newest technology products, and the demands of the work advance as technology progresses. Inductoheat's viability comes from proprietary experience in building specialized equipment; an inductor, for instance, may be a "work of art," not built after only a few hours watching others do it. However, the specialists work elsewhere as required, which greatly simplifies capacity planning. Unless they overload total labor capacity, they will probably meet schedule.

In the process of workplace organization, four special departments were folded into others, and workers became responsible for their own quality. The quality control specialist now on board spends much of his time instructing everyone else. Inspection is performed only on incoming material. Another quality check is performed by assemblers who check the matchup of mechanical detail as part of staging material for assembly. In both fabrication and assembly, the nonrepetitive work does not much lend itself to control charting, so most of the quality improvement consists of refining the processes of both design and shop work. CAD/CAM linkage should reduce some of the errors, but the firm has not yet made great progress with it.

Preventive maintenance has not been totally developed, but workers primarily look after that also. They are responsible for the proper functioning of *their* equipment.

This egalitarian approach to shop work has been tough on supervisory and managerial egos. Supervisors and managers became facilitators, making it possible for hands-on people to get their work done. Supervisors coordinate activity, authorize overtime and hiring, but sometimes only instigate decisions by workers on such things as what kind of read-out devices to put on gauges. They are measured by (1) quality of performance and progress of the work force and (2) making due dates. This has been a loss of power in the sense of ordering people around, and it takes acclimation to accept being measured on what almost everyone else does, not you.

Gone are any measures resembling labor efficiency or equipment utilization. They are unnecessary. People and equipment generally perform when needed.

There is no union. Pay increases are based only on merit. Wages and salaries are discussed openly. Cost-of-living increases or "automatic" increases of any kind are a thing of the past. No job guarantees exist, but job security is "understood" after five years of merit work and as long as merit work continues thereafter. One veteran was dismissed after 10 years because of inability to regularly arrive on time for work with his team, and it is considered important to the morale of those giving extra effort not to mollycoddle such cases. There was no special problem in reducing the work force through attrition.

Profit sharing sweetens everyone's pay. A small quarterly bonus is based on percent of gross earnings. In addition, an annual bonus is declared, usually 15 percent of gross profit or 15 percent of gross wages, whichever is higher. The effect is heightened interest in waste elimination and tolerance for the Spartan appointments throughout shop and offices. Everyone would rather spend the money on something else. Shop workers do not restrain criticism if new carpet is ordered for an office.

How did Inductoheat start all this? It got a general direction in 1982 and started reorganizing work in the shop, learning and planning the next moves as it went. It had no detailed five-year plan (and still does not). Too much planning is regarded as a waste also. The next moves are mapped out no more than six months to a year in advance. For Inductoheat, the world is too uncertain to project operations changes much further than that, but no one can call what it does only "muddling through."

COMMENTARY ON WORKPLACE ORGANIZATION

Inductoheat is a lean and busy organization, not eager to host parades through the plant or play part-time consultant. Characterize it as short on talk and long on doing, but it will share ideas with other doers.

Larger manufacturers doubtless will find the Inductoheat approach a little more extemporaneous than can be carried out in a larger organization, but the principles in its story are useful for anyone. The description emphasized workplace organization, but everything is connected to everything else in JIT, and therefore workplace organization is part of a total fabric.

Workplace organization can be done by any manufacturer—if management has the will. A common denominator of top-quality Japanese job shops (not all are in that category) is good workplace organization, what Japanese refer to as either the 4S or 5S approach. This is almost always presented to visitors as part of the management approach, but too often it is presented as "housekeeping," which implies only cosmetics, and as simply part of everyone's general understanding of how manufacturing should be run. Unfortunately, it is not part of everyone's general understanding. Where it is, the JIT approach is easier to accept. Workplace organization is part of the bedrock of that understanding.

Many of the problems are not with the workers. Once they understand, worker acceptance comes easily from most. (Acceptance from a political union leadership may be a different story.) The psychological hurt is in the ranks of management, staff, and supervisors. In all likelihood, the new roles of staff and managers will not coincide with the views that many have of their positions in life. Some typical remarks:

> I'm not really doing anything; there's no recognition here anymore. (Translation: There is a shortage of crises, and I do not know how to behave without one.)
>
> It could be done quicker if we put some experts on it. (Translation: I do not get to show my expertise or be fully in charge of a project. It is no fun being an advisor just explaining things.)

Learning how to work *directly* with hands-on people is difficult for staff people, and learning how to work with them *differently* is difficult for many supervisors. However, a major problem for all of us is acceptance of responsibility when a visibility indicator of nonperformance points to us, and it is certainly one of the major reasons why JIT/TQ philosophy is easy to preach but hard to do.

CHAPTER FIVE

Attaining the Effect of Automation without the Expense

Why automate? Because when done well, automation attains much more than merely substituting machine labor for manual. The objective is to make overall improvement in manufacturing—to eliminate waste, improve quality, reduce cost, and increase flexibility (the same goals as for manufacturing excellence).

Automation affects the total company. Many organizational functions must be closely integrated, starting with product engineering intertwined with production process engineering. A product designed for robotic assembly needs parts that can be grasped and oriented by robots. Seams welded by robots may be different in design than those welded by a human. A human can compensate for seam flaws, but a robot can't. However, a robot can reach angles very fatiguing to the human.

The engineering of automated processes must create the exact sequence of operations and the positioning of material. It must be particular on work envelopes, reproducibility of conditions, and possible exceptions. Variance of process capabilities must be kept narrow enough to prevent random defects.

However, this is not enough. As we have seen, quality problems can come from anywhere, and one of the more common problems observed in automated machining cells, for instance, is finding them choked with cracked or porous castings. Sometimes they are equipped with automatic sensing to detect these conditions and also to compensate for dimensional variance in the castings, which should not exist in the first place.

Design of the total automated process must consider quality in as well as quality out, product design evolution, maintenance, training, versatility of operators (or of programmed equipment), schedule change flexibility, changeover times, scheduling methods, and on and on. Handling and turnaround of tools is as important as that of the material. Lead times are important. Putting all this together, the conclusion is soon reached that a company must develop the way automation will fit with the total organization and the way the organization will fit with automation: marketing, personnel, scheduling, purchasing—everything, not just engineering and production.

If a major automation project is not considered organizational conversion for an entire company, the likely upshot is that engineers will oversell the technology while detractors will hold back because they do not see how to change all the aspects of management that will be affected. Each functional specialty will see it its own way. The pieces may not fit together, and people are unprepared.

Distraction is easy because automation makes an impressive show; robots and flashing lights suggest a technical fix without the necessary organizational development. Automation may not look so impressive in the financial statements. Video tapes of unmanned operations are taken while the process is running and technicians are nowhere to be seen. Overhead is tucked away in less photogenic locations. It is useful to remember that while those factories that run untended all night represent remarkable attainments in defect-free operation, the early ones were developed by machine tool companies and feature much of the same equipment they sell.

Not that automation is bad. It depends on its cost and how it is employed. The evil is in being tricked into a false show, somewhat like a young man buying a truck and wanting the fastest, biggest, loudest, and shiniest. A well-built plain one would do the job with less cost and trouble. The rest is waste except in the owner's aesthetic perceptions.

In the same way, watching a robot perform a task that is really doubling-handling material (and thus should not be done at all) is painful. Equally wasteful are guided vehicles or even conveyors going round and round until the function of transporting seems lost in the tizzy of being a moving warehouse without observable destination. A great deal of showmanship consists of energetic, but unnecessary motion.

Much of automation's desired effect comes from the advanced preparation, and more from the discipline to use it wisely once in place. If much of the benefit can be obtained from preparation and discipline, perhaps it can be gained without investing much in the equipment and software itself. That, of course, is the idea.

GAINING DISCIPLINE

Spending big money quickly on automation is not wise. In the end, its effectiveness depends as much on organizational preparation as on money and technical prowess. Tough, equally skilled competitors cannot be beaten just by outspending them. Major automation cannot be effectively "installed"; it must be accompanied by a way of organizational life.

Most companies do not have the capital for heavy investment in automation anyway. Of necessity they must try to capture as much benefit as possible with minimum expenditure. Even the company with excess cash does not wish to squander it. Study existing operations and remove the waste first. Prepare the organization for automation. The race goes not to the first to install robots but to the best overall effect. Automate only the necessary, value-adding operations and eliminate the rest.

A company whose access to capital was limited by its financial arrangements, such as Harley-Davidson, cannot spend much on capital equipment. Its approach is to work on the process and develop the people. Then, target automation so it contributes to elimination of waste and improvement of quality without adding overhead. It is a pay-as-you-go approach.

Equipment does not have to become worse with age. In fact, properly modified and maintained, it should become better able to perform its manufacturing mission than when it was new. If the equipment was designed to be maintained, wear and tear should be correctable; and, if the machine is carefully studied with respect to its mission, modifications to fit the purpose should make it perform better than when new.

An entrepreneur starting a new manufacturing business can seldom do so with all new equipment. The necessary investment base would be too high. The competition is using a mix of equipment, some new, some already written off, and if all new equipment is selected, its advantage over old must be great.

Skill makes the difference. There is no substitute for understanding the equipment, automated or not—how to operate it, how to maintain it, how to modify it in small ways for specific tasks. Computerized equipment requires computer "jockeys" in the same way that other equipment requires operators and maintenance personnel who understand it. They understand the quirks of specific machines in specific locations assigned to specific manufacturing missions, just as over time the driver and mechanics assigned to a specific car on a racing team become one with the machine.

No one gets a major automation project down right the first time. Even if they do, conditions change—product lines, part volumes, part configurations, etc. Therefore, flexibility is important.

Flexibility is attained in several ways, starting with volume flexibility. A very smart idea is to design a plant to have a very low break-even point so as not to create a monster that demands to be fed.

A second way is to incorporate general equipment as much as possible. Special equipment—designed for a specific part—is useless if the part is no longer used by your company or by another to whom the machine can be sold. Design floors and locate utilities so machines are movable without breaking the treasury.

A third way to achieve flexibility is to avoid great reliance on technical adventure. New technology involves risk, and some must be taken to make progress. However, attempting technical advance on a broad front at one time invites more debug than time and energy can fulfill. Try it a piece at a time if possible. The purpose of a factory is reliable, low-waste production, and a great one does not have to be state of the art in all technology at once.

Finally, flexibility is bounded. No factory can do everything equally well. *A very important decision is the mission statement for production:* What shall be produced. What shall be purchased. What range of work shall be encompassed. There is a limit, and stating clearly what a plant will prepare itself to do crystalizes effort. By exclusion, it will not fritter effort by attempting too much. For instance, a common problem in mature companies is attempting to make spare parts for ancient products in the same areas used for current production. The original equipment, tooling, and know-how are no longer present, so define a secondary area for such parts or farm them out. (It sometimes is a great business for retirees.)

Setup Time Reduction

Setup times are frequently cut by 50 percent or more just by being organized for it, that is, having everything in preparation when time comes for the setup to be executed. (*Setup time* is the downtime of the operation to change from one part or product to another; but, to reduce setup times, one should also remove the waste from the setup process by reducing the labor required for them.)

Setup time reduction relates to everything else. An example from an actual press department forming some simple metal parts is easily understood. Times have been reduced from an hour down to 10 minutes, and they are shooting for 5. First they found that die adjustment took a long time because gauge variation in the steel was high and so they had to tweak to compensate. Someone had found a good deal on steel coils. Improved steel cost more but was easily paid for by scrap reduction and the increased time between die maintenance. Then they assigned specific dies to specific presses, located them for easy access, and made other revisions to decrease the physical work and time for die exchange. They refined their preventive maintenance on both dies and presses. Now they are working on establishing a daily schedule for regular times to set up, and they are working on standardizing die heights to further reduce adjustment time. Everybody learned a lot and continues to do so.

Over time, a 90 percent reduction in setup times is not unreasonable. Half or more comes from workplace organization making ready for readiness for setup change as soon as the machine stops. Think of setup as a standard activity to be performed as a routine—the difference between changing a tire on a passenger car for the first time in five years and changing one in a pit stop during an auto race.

One common problem is to think of machines independently for setup time reduction, so only one or two machines are selected for trial. That does little good if the operations feeding those machines still have large lot sizes and if output is demanded of them in large lot sizes. Think of reducing setup times through a complete flow of production.

Another problem is not having a clear mission for the machine. Technicians and operators will concoct many inexpensive ideas for setup time reduction once they understand it is important and once the problem is clarified, but this cannot be done if they cannot anticipate what part is to be produced on which machine. The issues revert to quality, maintenance, schedule stability—and to workplace organization.

Most common is that tooling and tooling maintenance must be shaped up to permit fast setups. If deficient, it will at first be seen as an overwhelming obstacle. If tooling is so poorly designed or so worn that every setup must make compensating adjustments, refurbishment of tooling can be so time consuming and expensive that setup time reduction will be a long time coming. *Go for quality in setup first.*

Study of setups goes to the heart of production skills. A big difference exists between having just enough skill (and desire) to patch together tooling and equipment and developing the skill for defect-free, quick-setup production. Usually, technicians and operators must go through a process of self-improvement and self-discovery, and the equipment cannot be reconstituted any faster than the people who must work with it.

Once tooling comes up to snuff and workplace organization is under way, ideas for setup time reduction by modifying equipment begin to pop. Other ideas to eliminate adjustment time are generated. Try to make the first piece produced a good one. Locate gauges and instruments nearby so those doing the setup can check the first piece immediately.

If current practice is to check the first piece by an inspector, an organizational obstacle must be breached. Focus on how to enable operators or others to assume this responsibility, not on the reasons why that should not be done. Underlying the whole effort is a process of skill and confidence development.

If possible, locate gauges and instruments so that long, time-consuming walks to check parts are eliminated. The author visited a press shop in which operators carried the parts to a distant quality office for a gauge check, then waited for the results. Each trip averaged two hours. A full setup should have taken 10 minutes or less if everything were developed for it. On another occasion, a machine operator was trying to set up a machine to a one-thousandth tolerance using a six-inch rule, then taking a trial piece to a distant optical gauge for verification.

Practice and timing are important. If operators and technicians set up the same machine and tooling for the same part numbers every week, they soon become good at it, and being good at it requires a standard procedure. If they are constantly surprised by schedule changes, they soon become discouraged. Timing and repetition allow everyone to get setup down pat. Refine it using cycle time analysis.

Do not turn setup time reduction over to manufacturing engineers. They will make a specialized problem of it and probably spend too much money doing so. Reserve their talent for where it is really needed, and the true engineering challenges should keep them busy. The operators and technicians need to develop. That is part of the purpose, and besides, they are likely to develop a succession of simple, inexpensive ideas that will accumulate to a big effect over time. Not all ideas will work. When they fail, try again. That is part of the learning—as long as it is not overly expensive.

A great deal has been written about setup time reduction, and nuts-and-bolts ideas abound. One should not throw up the hands, saying that a complex machine tool is an impossible challenge. It is more difficult than a small press, and perhaps a setup time in under 10 minutes is not possible, but make an effort anyway. Mind over matter accomplishes a great deal.

Layout for Improvement

One of the first ways to begin capturing the "effect of automation" is to change the layout to minimize travel distances of material, tooling, and people. Many plants assume that they must operate by a job shop layout, with all equipment and operations of the same type located together in the same department. The problem of locating job shop departments to minimize travel distances has been studied to death. The big gains come from questioning this assumption and locating operations in the sequence in which they are performed on material.

This forces a rethinking of how work is done and how layouts are constructed in detail. Some of the issues are: common routings for families of parts, exceptions to those, and the reasons; dirt and contamination, such as from heat-treating and painting operations;

noise and vibration; location of tooling and gauges or other instruments.

Those are the physical problems. The organizational problem is that the advantage of locating people of similar skills in one area will be lost. If departments are broken up, the issues of special skills and training must be faced squarely. Often these are not as great as was feared, and they are offset by the advantage of immediate feedback and visibility of total operations.

The core problem may well be the status systems built up by isolated specialists in different departments. Breaking these systems is difficult even if people can readily observe that skill maintenance problems can be whipped.

Another tough problem is equipment selection and modification over a long period of time. Equipment intended to be used in-line will be selected if easily modified to balance the line in which it is currently placed. If it is located in a department of similar equipment, thinking of it as a stop on a "standard routing" will be less likely. Try to clarify the specific mission of equipment by in-line layout, if possible; but if not, think of equipment as part of a flow anyway—except as widely separated stops in a flow. Develop a layout for standard conditions before exceptions.

All layouts have compromises, but try as best as possible to:

- Simplify flows. Material flow one way; tooling in a cross-flow. Tooling is sometimes a much bigger material handling problem than material.
- Minimize material handling, and keep it simple. Avoid lifting, by worker or by machine. Keep lift trucks out of the operation's core. Close up space between equipment.
- Make use of people 100 percent. Provide a layout that promotes visibility and flexibility. Ability to respond is a big human advantage.

Figure 5–1 sketches several different layouts that promote flexibility. A *variable* work cycle can be increased or decreased by adding or subtracting people and promotes their trading work between them.

Rather than attempting to balance work between people so that all share an equal load but most are underloaded (bottleneck), try to shift work so that all work at a full pace save one. Pick a skillful one to be off-loaded. That one will likely find a way to further shift

FIGURE 5–1 Flexible Layouts for Variable Work Cycles

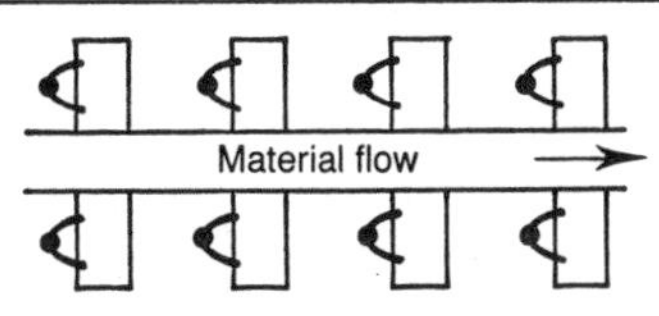

Bad: Operators caged. No chance to trade elements of work between them.
(subassembly line layout common in American plants)

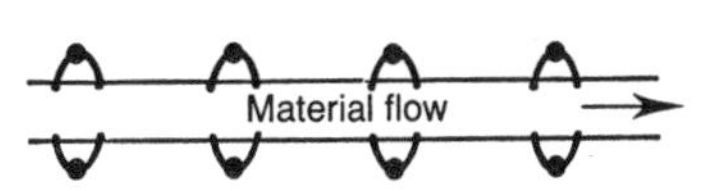

Better: Operators can trade elements of work. Can add and subtract operators. Trained ones can nearly self-balance at different output rates.

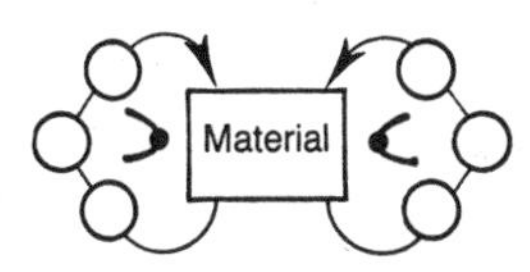

Bad: Operators birdcaged. No chance to increase output with a third operator.

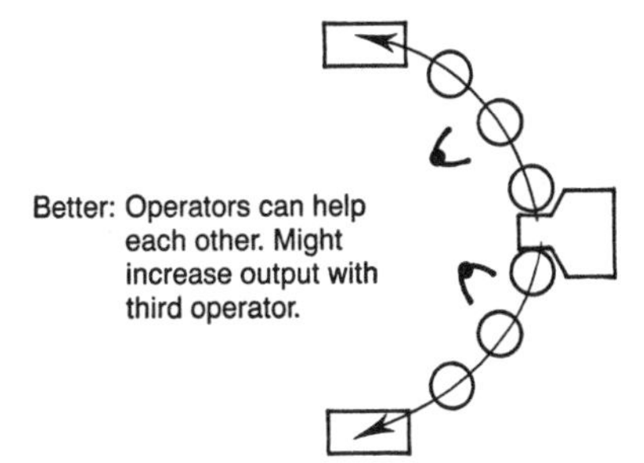

Better: Operators can help each other. Might increase output with third operator.

Bad: Straight line difficult to balance.

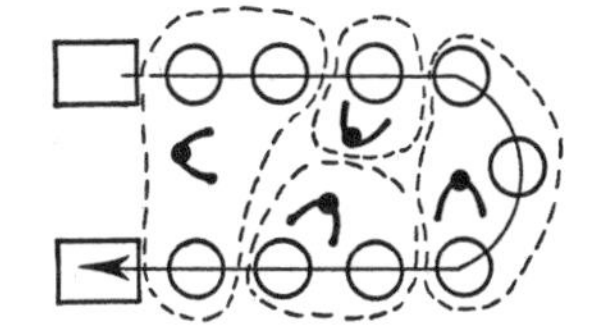

Better: One of several advantages of U-Line is better operator access. Here, five operators were reduced to four.

or reduce work so as to become unnecessary (once the idea is understood) and therefore useful on another job.

Many simple layout ideas contribute to eliminating waste. A half-step here, a reach avoided there, and a double-handling eliminated somewhere else begin to add up. Layouts also contribute to JIT. Much of what passes for miraculous scheduling is actually a cumulation of skillful work developing people and layout to do what is necessary when necessary; and much of the preparation of people for JIT is teaching them how to improve their work methods and layout to handle the problems of mixed model assembly, for instance.

Suppose that one could concoct a wondrously fast machine to automatically insert all the components in hundreds of printed circuit boards used in dozens of end products. Getting all the material and tooling into the machine and away from it would be quite a problem. Computer systems might keep track of thousands of parts, but the physical layout would be exceedingly complex. The point of this exaggeration is that layout complications in production (not just automation) should be a warning that the fundamental concept of the process is flawed.

A great deal of automation is only moving something. An ingenious layout goes far to prevent automating a great deal of motion that should really be unnecessary. *The value of intense effort integrating operations into an effective, low-waste layout is greatly underestimated.*

In a well-developed JIT operation, little is said about the problems of scheduling. Instead, much is observed about adjusting work and layouts in order to make any "reasonable" schedule possible to run.

Observe how much space is taken for operations that do not add value directly to production. Offices encroach on production space. Room is taken for material handling, quality inspection, rework, and storage. It is not unusual for a plant covering thousands of square feet to use only 15 to 20 percent of the space for first-time material conversion.

Even in the direct production areas, spacing between machines is often left wide to accommodate various problems. Extra space in a plant is like extra space in a garage at home. It will soon be occupied with something.

By concentrating on workplace organization, quality improvement, and common flow paths, most plants can open up space no one realized it had—often 10 to 20 percent in short order, and up to 50 percent longer term.

Much of automation's intent is to capture the precise location and orientation of tools and material and maintain them without exception throughout production. Another major objective of automation is to perform various quality checks without losing locational control of material. If these objectives can be attained in simple, inexpensive ways, much of automation's effect is obtained anyway—by developing detailed improvements in layouts, quality, and methods of work.

The capital savings from space reduction come once. The operating cost savings go on and on and include avoidance of wasteful automation. They derive from reduction in material handling, better visibility, and prompt attention to problems. Excellent layout is integration of operations in detail.

Office planning is also a factor. If office space is not provided, excess staff is hard to accumulate. Likewise, offices close to the operations promote visibility. It is very easy for managers and staff to work under a plant roof but rarely see the production operations themselves or, if they do, only a minute portion. If visibility is important for workers and supervisors, it is important for managers and staff also.

The Cell

One of the best ways to reduce material handling is to perform as much value-added work as possible in one spot. Multiple operations can take place on one machine, for example, multistage (progressive) dies in one press and multiaxis machine tools. However, the added complexity of doing it all on one machine may add significantly to setup time, and both the machines and the tooling for them may be expensive. However, if technically and economically feasible, doing it all on one machine is a good idea.

A cell of machines in which parts move one at a time from machine to machine is better for some purposes. Each machine may be easier to set up, the overall cost lower, and the cycle time working on each part shorter. As a simple example, consider a five-axis machine that will perform all the operations that five simple machines will do. Suppose the five-axis machine costs $1 million and takes two hours to set up. The five simple machines cost $100,000 each and require 10 minutes each to set up. Combined into a cell, they process parts in one third the time. Even if the cell takes an extra operator, the choice is obvious if the same quality results and the same maintenance time is required. That is the basic idea and advantage of a cell. This example was oversimplified, but grouping operations into cells is an idea worth considerable investigation.

A cell generally requires more versatile, trained operators. A history of specialized operators can be a sticking point, but the overall idea is to develop people, so work toward diversification.

Some of the machines in a cell are probably not used to their capacity, which is a concern if machines are expensive and a reason to avoid expensive equipment when possible. It is not use of individual machines but rather the overall effect that counts. The overall effect is to (1) greatly shorten lead times of material through production and (2) provide excellent visibility and immediate feedback among the operations in the cell.

Why is so simple an idea not widely used? Unless the idea of cell production is in the mind from the beginning, conversion to it seems difficult, if not impossible. Product design and process development have not worked toward it. Operations have been developed independently.

Individual engineers design one part at a time without much reference to how the part fits into a family of parts, all of which may be made in almost the same way. Machines are selected on independent merits. Some run fast; some slow. Some are easily maintained; others have considerable downtime. A different mix of parts has been assigned to each one.

Then, equipment proposed for a cell is used for items that do not go through the cell. The interference destroys smooth flow. A breakdown on one machine hangs up production for all. Piece-at-a-time transfer of parts between machines that run at different rates is possible but results in a high percentage of idle time for the fastest machines.

These problems may or may not be technically surmountable. Those in the mindset are the hardest. One is the urge to run a machine at its fastest rate even if the volume of parts needed from it is so low that the machine is otherwise idle most of the time. The cost system presumes independent operations, so the cost for running machines in a cell will appear to sum to a higher cost than if they were run independently due to multiplied downtime and so on. The overhead cost required to run independently is fuzzy to estimate. Continuing to route production as seems best for each part is easier for manufacturing engineers in the short run, and design engineers can continue working independently if they have little discipline for standardization.

Those companies with available technical firepower are inclined to engineer through all of this. Redesign it, computerize it, simulate it, and start over, which is very good provided the cost is not too high and the original mindset is not still there waiting to hobble the technical achievement.

Evolution to a cell may work better, provided the equipment is simple enough to do it. This route has two major advantages. First, the experimentation is with the real thing, so problems of quality, layout, and work flow are worked out directly, debugging your way through it as you go. Second, chances are good not to make a slip up that prevents flexible use of the cell.

A very simple way to connect equipment with fixed work cycles so as to have a variable rate of output is shown in Table 5–1 and Figure 5–2. (This is the principle behind all the homegrown automation in Toyota's Kamigo engine plants.) The cumulative man-machine chart is a basic version of the kinds of analysis methods used for thorough cycle time analysis (Chapter 4). Continuously working away at cycle time analysis wrings the waste out of these operations.

This approach is the fundamental method for carefully studying total operations *before* spending big money on equipment. Needless to say, the entire work force must be developed for it. They must learn to think continuous improvement continuously. After they do, they are tough to beat with a straight engineering approach.

U-Lines

A simple cell has equipment tightly arranged in a U-shape within which workpieces move one at a time from station to station around the U, usually by one or more workers inside the cell. This is useful for parts small enough to easily hand-carry and when using simple machines small enough that operators have easy access. Most equipment can be modified to go through a processing cycle without the operator watching. A large U-Line might have 30 machines and two or three operators. The rate of production can be adjusted by increasing and decreasing the number of operators; and, if equipment is simple, the operators can change over the entire cell to another part in 15 minutes or so.

The first technical throught is, why not move the parts by robot? Perhaps that can be done, but whether it is useful is another consideration. If a robot were used, would a human have to be present to monitor the cell anyway? Would a technician need to give full attention to keep the equipment running? An operator "walking the beat" inside the cell performs the monitoring function, constantly observing both parts and equipment, close up and firsthand. And this kind of cell can be developed with simple modifications

TABLE 5–1 "Pull System" Logic Rules for Automatic Transfer of Workpieces between Machines in a Cell

Think of workpieces going through just one machine in a series of machines:

Straight Flow, Identical Workpieces:	*Complex Flow: Multiple Workpieces (Some workpieces skip some machines; different workpieces have different operations from same machine)*
1. Sense workpiece in load position.	1. Sense workpiece in load position. *a.* Identify workpiece. *b.* This workpiece to this machine? *c.* If not, skip. If so: (1) Download correct program. (2) Select correct tool.
2. Load workpiece. Perform any failsafe checks on previous operations by sensors.	2. Load workpiece. Perform correct failsafe check on correct workpiece by sensors. *a.* Load tool, if necessary.
3. Work cycle.	3. Work cycle.
4. *Idle* until previous workpiece empties from unload position (is demanded by next machine), then eject.	4. *Idle* until previous workpiece empties from unload position (is demanded by next machine), then eject.
5. *Idle* until next workpiece arrives.	5. *Idle* until next workpiece arrives.

Several lessons come from reflectively examining these five steps:

1. Straight flow could be mostly automated with microswitches. Even the failsafe sensing can be very simple if it only senses for presence of a particular characteristic of the workpiece.
2. Synchronization between operations is achieved by allowing machines to sit idle between cycles.
3. In automating:
   a. *Straight flow:* Simple automation. No information must be carried by the workpiece itself.
   b. *Complex flow:* Potential setup change for each workpiece, so total work cycle equals setup cycle plus run cycle. The machine must interpret information carried by the workpiece (standard routing).
   c. *Non standard flow:* Information must be carried by the workpiece, in some fashion, telling which machine to go to and what should be done to it there (job shop case).

Conclusion: The closer operations can be organized for straight flow, the simpler the automation. Complex automation is complex logic and complex layout.

FIGURE 5–2 Flexible Output from Fixed Work Cycles

Fixed Work Cycle Machines In Tandem

Automatic feed between machines by the method of Table 5-1 allows equipment to be idle between cycles. Total cell output rate can be controlled to any level limited by the longest work cycle time (machine 1 in this illustration).

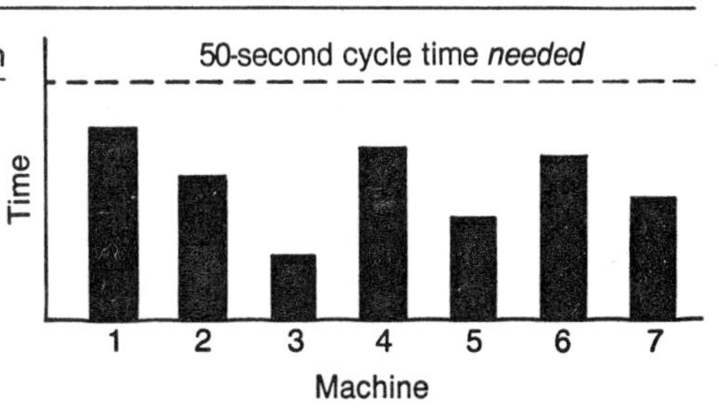

Man-Machine Cell (May Be U-Line)

Cell output rate is controlled by variable work rate of operators, not the fixed work cycle of any machine. Limit is longest man-machine cycle (machine 1 again). If machines are simple and inexpensive, start another cell with another operator if more output is desired.

Cumulative operator time per cycle is 88 seconds, so two operators are needed to meet the 50-second cycle.

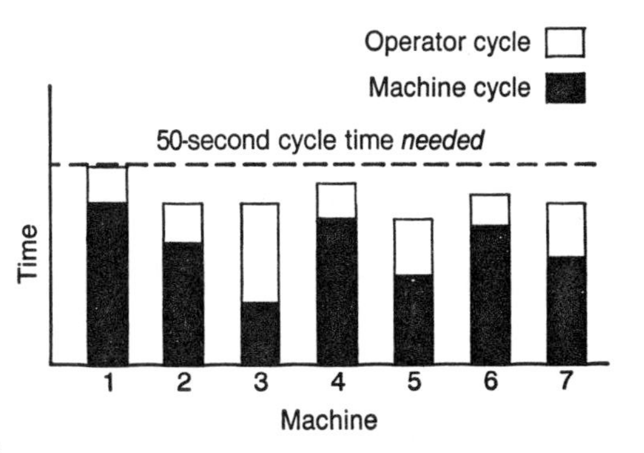

Cumulative Man-Machine Chart

The chart shows the interaction of the two operators with the seven machines. Variable operator cycles control output rate.

The chart itself can be done in much greater detail, and it is an excellent tool for cycle time analysis (Chapter 4).

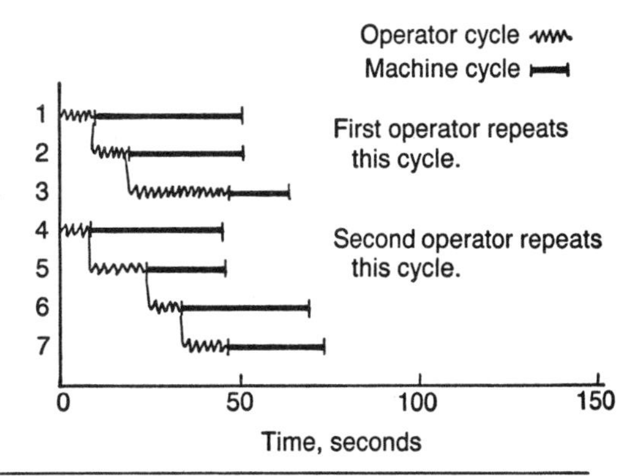

of simple equipment—no complex sensor systems to maintain and no software to debug.

Will a U-Line always be superior to full automation? Of course not. The point is to evolve at low cost toward the desired overall effect.

Flexible Manufacturing Systems

The term *flexible manufacturing system* designates a system of machines under centralized computer control served by automated computer-controlled material handling equipment. Workpieces may move in sequence from machine to machine, or work may simply be assigned to the necessary machine when it is open. In practice, an FMS generally consists of a computer-controlled cell of CNC (computer numerically controlled) machining equipment with its associated load/unload equipment and an automated storage-and-retrieval system for material and sometimes for tools. Often the workpieces are too heavy for human lifting.

Sometimes flexible machining cells work well; sometimes not. The difference comes down to whether the company can overcome the same kinds of technical and human issues associated with other kinds of cells. Fundamentals of people preparation, quality, layout, maintenance, and so forth make the difference. Both computerization and the price tag make the FMS high profile in management attention, but the success factors are buried back in low-profile activity.

Think of the Entire Plant as a Cell

This is one way to stimulate thinking about operations as a whole and not piecemeal. Operating an entire plant using workplace organization and high-visibility practices creates that overall effect. It may not be quite as closely meshed as material moving through a set of machines organized into a cell, but the results are similar.

The objectives of revising an entire production facility are the same as for creating any individual cell: minimum travel distance, high visibility, and immediate feedback. This suggests the overall effect desired. The considerations in obtaining it are similar; the problems much the same except the entire plant is a bigger arena. Can the product be designed with families of parts having similar operations in production? Can the operations for existing designs be organized into a standard sequence? Can such problems as a central heat treat facility be worked around in some way? Operators and support personnel developed? The schedule stabilized into a repetitive pattern for it?

At one company with an 18-person operation, engineers figured out how to accomplish the same tasks with six operators and four robots. They were busy trying to justify four robots on the basis of decreased labor for 12 people. After considering how they might further improve the work using cell arrangements, they found they could do it all with four operators and no robots. Now *that* is the idea.

The essential pattern of thinking is to stop considering operations independently and start thinking from the very inception of product concept how operations might all fit together. From an upper management viewpoint, this comes down to defining in a thoughtful way what the manufacturing mission of a plant should be. If a cell cannot be developed with clarity in mission, neither can a plant be. Missions inevitably change with the evolution of markets and products, so this is not a one-time decision for the ongoing business.

Performance Measurement for Plants and Equipment

The major problem with most existing performance measures is that they assume operations are independent. Take equipment utilization, for instance. If the utilization of every piece of equipment is measured independently, and if managers are seriously asked to keep the utilization high, the motivation is to run every active piece of equipment as much as possible whether production is needed or not.

The actual measurement is a lesser problem than how it is used to motivate performance. If production supervisors are driven by it, they do not want to shut down equipment until they have to (meaning that they can find no more work to put on it). If a machine is running well, they let it run as long as machine, tooling, and material hold out. They assign to each machine the operator who is most likely to maximize output from it and reassign them reluctantly. (Labor efficiency motivates in a similar way.)

This motivation is the opposite of almost everything desirable. Managers so motivated cannot put heart and soul into making frequent setup changes, nor will they take seriously a policy of shutting down at regular times for tool changes, quality checks, or preventive maintenance.

Harley-Davidson adopted the correct solution for this problem. Management threw the equipment utilization report into the trash can and instead measures equipment availability. Will the equipment do what it is supposed to do when needed? There is no point in running equipment just to run it.

Measuring equipment availability stimulates preventive maintenance and is consistent with process improvement and cross-training of operators. (One major reason for equipment downtime is that no operator capable of running it is available.)

Other performance measures must be suggested, with the caveat that all such things are subject to the mission decided for a production facility. Here are some favorites:

1. Number of setups per month for particular pieces of equipment or for departments. The objective is not to maximize this but to hit a target that makes sense for the current schedule and the state of development for flow production and quality. One also cannot exceed the total labor available to make setups.
2. Process capability. Is process capability within the necessary range? Number of operations failsafed where desirable.
3. Flexibility. Number of operators capable of operating each piece of equipment, for example. Number of technicians capable of servicing each one.

One can measure to death and still not know much. Workplace organization is judged by seeing it. Selecting a few measures compatible with heading in the desired direction is useful, but hundreds of them add to the data collection effort without contributing much except the waste of collection.

The Virtues of Inexpensive Equipment

Inexpensive equipment comes with little obligation to use it. Managers do not strain to obtain an ROI or a payback from it as an individual piece. The thing is just there, with all the other equipment, to be used or not as need be.

The most important part of tooling and equipment—no matter how automated and advanced—is the people associated with it. American combat flying is considered superior because of both superior training and superior equipment; but, of the two factors, superior training and experience of air crews and support crews are the most important. The best crews show well even if their equipment lacks the latest technology. There is no reason to think otherwise of production personnel and their equipment.

Defect-free production using no-frills tooling and equipment is achieved through people skill. By it, equipment is adapted to new missions, and experience is formed on what is really important in assembling the tooling and equipment for new work. By working first to upgrade and modify equipment already owned, expenditure for more is minimized.

With equipment and tooling cost at minimum, break-even points are low, which decreases risk in a business downturn. It also reduces incentive to keep equipment running so that its amortization can be spread over more units of production, which hopefully will be needed in the future. Gambling to pay off a high-stakes equipment investment with a higher stake in inventory buildup is an adventuresome risk.

Flexibility is derived from the attributes of the equipment, tooling, and people, and it is promoted by low investment, but smart investment, trying to stay flexible so as to quickly put together an arrangement to produce what is needed on short notice. An assortment of simple equipment that can be quickly rearranged within a U-line cell is a prime example.

Inexpensive equipment can be held in reserve for substitution while modifications and improvements are made to on-line counterparts. The policy of assigning the same parts to the same machine seems in conflict with this (and partly is) because changing machines is one more source of quality variance to deal with. However, there is a difference between planned substitutions for improvement and haphazard changes for whims and emergencies.

Also, if no pressure is exerted to run equipment full out, different pieces of equipment can be made to run at matching speeds. The faster pieces can have a delay added to each cycle to slow them to the speed of the slower pieces; or, even better, the complete series of equipment can be adjusted to run at the rate production is needed. Idle time on equipment is accumulated that way but, if the output is not needed, so what?

Inexpensive equipment is easier to regard as part of a set of equipment. Inexpensive operations are easier to regard as a unified whole, not as a collection of independent activities, some scrutinized to death and some virtually ignored.

Operations are not independent, but management thinking often assumes they are. Plant efficiency is assumed to be the sum of efficiencies of independent operations, but if operations are not independent, the criterion that counts is whether the plant as a whole made what was needed without error, on time, and at low cost.

Accounting models state that aggregate costs are the sum of independent costs, which is true for the model, but it does not follow that the operations on which the costs are based are the same as the model.

PREVENTIVE MAINTENANCE

An assumption too often made is that equipment must become less fit for use as it ages. Its purchase cost has depreciated to salvage value, so the machine must have done likewise. Unless a machine was poorly designed for maintenance from the start, that assumpion is not true. If worn parts can be replaced, the machine should not only be as good as new but better.

Most machines are used for a particular application. They are modified and continuously improved to perform that application better. If they are well used and maintained, they should be better for their purpose after 5 or 10 years than when they were new.

Obsolescence is a different issue. Perhaps a differently designed piece of equipment can perform the same task with less waste, or perhaps through technology the tasks for which the machine was originally developed are no longer needed—the process is obsolete. *Then* the machine is either junk or a museum piece, but not before.

Preventive maintenance (PM) is important to preserve the equipment and even more important to preserve quality. Many PM programs fade away bcause this concept is not properly understood. In a plant where emergency production always prevails, time for maintenance comes last. The machines most critically needed will not be shut down until maintenance is no longer preventive but instead necessary.

If preventive maintenance is understood only as an insurance program, it is analyzed as a trade-off. Is it worth $100 per week for three years to delay an overhaul for one year or to prevent a catastrophic malfunction at an unexpected time? If there is pressure to produce, even an answer in the affirmative will be deferred.

The real reason for preventive maintenance is to preserve the process capability of the equipment and tooling. If that is done, the equipment must also be preserved from excess wear and tear, and the standard of excellence to which each machine must be maintained will allow failsafe production of quality output. Preventive maintenance is one of the quality cycles to which many production operations are subjected. Oiling equipment once a week is inadequate PM.

Preventive maintenance applies to instruments and tooling as well as machines. A regular program to keep all instruments and gauges in calibration is one of the most important aspects of total quality. Without accurate measurement, the process capability of everything else is endangered. This is one of the most vital areas for preventive maintenance according to quality cycles appropriate to each instrument.

If this action has not already been taken when starting total quality, it should be one of the first. A sign of a well-run plant is an excellent visibility system to assure that readings from all measuring instruments are kept calibrated and that the same kinds of instruments are as identical as possible. (Inattention to calibration is one of the most common "black holes" in plant quality issues.)

Operator participation in equipment PM is highly desirable. Operators who help care for equipment understand it better, and they are less likely to abuse it. Those who clean see things missed by those who do not, and they become sensitive to any minute change in sound or motion that signifies a potential deviation from normal operation.

Maintenance also has a purpose other than regularly keeping equipment and tooling up to snuff. Those who maintain the equipment learn it thoroughly. They have more ideas on how to modify it for improvement. A plant with good internal capability to maintain and modify equipment is a leg up when it comes to revising equipment for setup time reduction, cell production, and other purposes. They can keep the equipment at the necessary level of process capability, and they can use it in a flexible manner.

Preventive maintenance should be one objective of a visibility system. Put cards or logs on machines so that operators can note suspicious functioning that they cannot tend to themselves. The logs also keep the maintenance history right on the machine, which helps. A way to note that a machine, tool, or instrument has been tended by maintenance is to put a colored sticker on it. An old sticker shows it has not received attention, and the stickers easily denote machines that need it, even from a distance.

One of the toughest problems is that preventive maintenance can become perfunctory. A technician who makes rounds of machines that seldom have anything wrong with them can lapse into dry running it—change the sticker without doing anything. Making PM a quality program prevents this. The technician must record something checkable, for instance, bearing wobble. Working to improve process capability and make the machines better instead of just keeping them going reduces the incentive to dry run.

A good test of total work force skill is to check what equipment in a plant can be disassembled and put back together without disturbing process capability. In fact, it might come out better. A skilled group can do a great deal, but this status is not quickly and easily attained.

As more and more computerized equipment is used, software maintenance becomes a major part of maintenance in general. Preventive maintenance is important for software also. As much waste proliferates from poor software as from poor hardware.

In addition to manufacturing support systems, software is directly associated with shop floor equipment—everything from numerical control tapes to programs downloaded from computers in far away places. Establishing regular quality cycles and PM activities is useful for software too. Visibility methods to patrol the current maintenance status of software are a little more subtle than with hardware, but possible. Preventive maintenance of data bases is generally considered measures for data accuracy.

Issues in software maintenance are well-known, but a brief review of a few of them holds great similarity to issues in PM for complex equipment:

- Documentation of current status is vital. The same is true of complex equipment. The best place for documentation is imbedded in the software itself. Same with equipment. Knowing what you see when viewing it directly is best. Separate manuals help, but they too need regular periods for PM—regular reissues. It helps to simplify them, standardize them, and have a regular time set aside to update them.
- Software often needs constant attention for enhancement and to keep current with hardware and interfaces. The same with complex equipment, and the more so if its use is intended to be flexible. Computer and software maintenance is mainly a holding action on obsolescence.

- Having multiple persons able to diagnose software quickly is important even if it is well documented. Production equipment of all kinds needs the same, and cross-training technicians is a problem that strains resources and taxes the ingenuity of management.
- Software is written in different languages in which not all programmers are expert. Complex equipment has electronics, software, hydraulics, mechanics—everything; thus learning the whole thing in detail is a challenge.
- Security is an issue. Software can be damaged or stolen, sometimes unintentionally. However, keeping it in a secure place far from the action does nothing to assist flexibility or immediate feedback—or maintenance and corrective action if visualizing the problem is helpful to the process, which may be true of shop floor software.

Little genius is needed to see that the potential for waste can multiply rapidly with complex equipment. In highly automated facilities, the common denominator is that a corps of technicians familiar with the complex equipment is the heart of the operation. Simplifying this situation reduces the potential for waste, much of which is rooted in complex maintenance of software and equipment truly understood by few.

PRODUCT DESIGN

Tremendous potential exists in designing a product to be easily made. A major automation project generally calls for a redesign of the product to standardize and simplify the parts. Few parts and simple geometry are easier handled by robots, and perhaps automatic fabrication and assembly is simplified.

However, after simplifying product design for automation, the same changes have sometimes made manual work so easy that automation was unnecessary—economically unjustifiable—in more than one case. This did not deter at least one company from going ahead with automation, however. Having set its mind on robots, it was determined to have them.

If redesign is of great benefit in simplifying and standardizing for automation, it is also beneficial if the actual automation never takes place. It helps promote production changes that provide the same benefit without the equipment. Rethinking the product design is part of studying the total manufacturing process before spending money on equipment.

Produceability is an oft-neglected aspect of product design, but it is only one consideration in design engineering. The product should also do what the user wants, be fit for use with no gremlins waiting to cause product liability problems, be durable, have an attractive appearance, and be maintainable without a total teardown. Most products also have to be transported to their place of use and sometimes stored and identified at various points along the way. Packaging is the most important part of some products. Finally, users must understand how to use it properly, so directions and manuals are also part of the total product consideration.

Put all this together, and much design engineering is a highly integrative activity. For many products, even technically advanced ones, only a small portion of the design work is nontextbook ideas. That is the exciting part for most engineers, but a great deal of their work is extension of well-known past practice to different applications. Much of this extension is integrating all these considerations into the specific physics, geometry, and procedures to be produced and marketed—and making sure it all is going to work. Design engineers are greatly maligned for concentrating so much on the technical aspects of their work that they give too little thought for how everything will fit together in their own organization and in the field.

To achieve manufacturing excellence, the entire responsibility of design engineering should be well done—total integration. Otherwise, production is a turmoil of conflict and engineering change. As we observed, issues relating to quality begin with the design concept. Converting this concept to a clear set of quality specifications is one responsibility of product design and development.

Begin with a clear set of goals for the product design from the viewpoint of the user. Product ideas may not originate in design engineering. Often they are the offspring of ideas coming from marketing or field service. Certainly the desired features of a new product in a large company are greatly influenced by the observations and data collected by marketing.

Specifications are confounded if the product concept is influenced by a number of parties who do not agree. Not much energy is left to make the product easy to produce.

One of the better design objectives ever observed was by a company competing with Xerox in the copier business. The objectives:

Make the image at least equal in quality and the operating speed at least the same. Make it half the size and to sell at half the cost.

Such objectives are challenging but clear. Whether they are met or not is easily seen, and the cost objective presses for a design easy to produce. It also creates much thought for low-cost methods of distribution and selling.

Had the objectives pressed for a major advance in image quality or duplicating speed, the produceability objective might not have been so dominant, but it should have been present and clear nonetheless. For instance, make it sell for no more than competitive price. If that objective is unstated, production methods tend to be ignored until technical ones are met, which will then necessitate some revising—the classic case of "throwing the design over the wall" to production.

Neglecting production considerations until late in the design work delays actual product introduction and is likely to result in a messy production start-up at best. The practice is not permissable in an environment in which cost is a driving factor and in which pricing is an aggressive bet on the outcome of a production learning curve. Design a product as far down a learning curve as possible.

Sometimes new production technology must be used to produce a newly developed product, but learning both a new product and a new production technology at the same time compounds problems. A wiser policy is to keep learning in bite-sized chunks by developing new products and new processes in different cycles to the extent possible. Design new products to fit established production processes, and change production technologies in anticipation of new product requirements.

The usual solution advanced for improving the produceability of designs is to move production engineers, purchasing representatives, and materials specialists further forward in the design process. Include them in the design team. If that is done, why not have the design team composed of marketing and distribution specialists also? Soon the design team becomes large and cumbersome, while a small group would enhance communication.

Nothing replaces having designers aware of many considerations of design without having to learn about them during a design project. Managing a large design team is harder than managing a small one, and if the number of meetings required exponentiates, the total design process stretches—not what anyone had in mind. Again, the best solution appears to involve the development of people. For manufacturing excellence, a multiskilled staff is as important as multifunctional workers.

Simplifying Design

One major way to reduce cost is to have fewer parts: fewer parts in total and fewer different kinds of parts. Forty percent fewer parts is less work in assembly, less work in fabrication, less work in finding sources for material, and fewer quality details to tend—provided the parts reduction is made without creating a complex wonder out of some of the parts so that degree of production difficulty offsets the reduction in part count.

A division of Tektronix found that it had a proliferation of different fasteners in its product designs. No one had ever told designers that standardization of fasteners was important. A committee formed to analyze the fasteners selected about 800, which they thought would serve almost all applications. These were mounted on a display board in design areas so a designer looking for a fastener had an easy reference. Selecting a fastener in this way is easier than every designer searching through all the variety in the world, and the result was a great reduction in the number of different parts in Tektronix designs.

This move greatly simplified purchasing of fasteners. It reduced the number of parts to be managed in Tektronix's materials system. It reduced inventory. Best, it reduced delays in production caused by lack of some small part or other.

Fasteners are seldom an integral item in the technical functioning of an electronics design. Parts custom-designed for product function are harder to standardize. *Group technology* attempts to solve this problem by classifying parts according to size, geometry, or function so designers can use existing parts instead of creating something new—and different.

In practice, group technology has had mixed success. Some designers may prefer their own unique imprint on the design. However, many problems come from the complexity of the design process itself. In the absence of a design policy, designers select few common parts because the intent of design is to fill a void in the product line—something different. Everything comes out "an eighth-inch off."

An example of a design policy for home appliances is to design each one around a family of standard motors and shafts. Then at least some of the most expensive parts will be standard.

If standard parts are not possible, try to simplify design using a protocol such as the UMASS system of design for assembly. Alternatively, programs in conjunction with computer-aided design can assist with design simplification.

In straight-flow production, if newly designed parts follow the same sequence of operations (or one that can be quickly set up), introduction is much smoother. Computer-aided process planning helps identify (but does not resolve) the "design traps" in which one can be cornered. Typical is a metal part that requires a hardening process for which a plant is not equipped. Not only might such a part require breaking up a cell in which current parts are induction-hardened as part of the flow, but the options are to get a new piece of equipment (and learn a new process) or send out the part for treatment. Either choice wins a raspberry for the designer.

Design traps can be minimized by a policy of designing all similar products to the established process—unless a conscious decision has been made to change the process. Naturally, exceptions will exist, but the policy coupled with design reviews can head off unnecessary problems. Establishing such a design policy presumes that the existing production operations sequence (or standard routing) is definable. This comes down to a chicken-and-egg question: Define basic process routings for designers first; otherwise, standard routings do not develop because designs do not call for them, and vice-versa.

Critics of this approach assume that it stifles creativity. It does. Not all creativity comes to a practical end. Much more is involved in innovating a new technology than in inventing the bare-bones working model. The rest is not just a matter of detail, as some technicians suppose, but the total process of refining it, producing it, educating users, persuading, and perhaps breaking old habits of thought. Technical progress does not come from technical chaos. It might start from chaos, but the rewards go to those who organize the total environment for it. Productivity should improve in engineering as well as in the rest of the organization.

As production processes and methods evolve, so do the design policies based on them—a symbiotic process. A company in pursuit of radical technology must take the risk of breaking the old shell that traps it in old process technology, but it must form a new one—or be beaten down the learning curve by competitors who do.

Structuring of Options

The basic problem is *variety* of product attractive to a *spectrum* of users while retaining operations as *simple* as possible throughout the company. Conventionally, if a basic design made to stock is modified for options, each modification is a different model. If a product is made to order, the customer specifies the options wanted. Options are assembled to order from predetermined parts, not custom-designed parts. Some options are special adjustment settings or special software.

Optioned products cover a wide spectrum: automobiles, machine tools, computers, office furniture, pumps—even commercial-grade door hardware. Structuring the options is a major part of making a product line competitive. Everyone would like to present customers with something they want (or think they want), but presenting them economically is how to stay in business while doing it.

Structuring options is an extension of design policy. Engineering policies desirable when designing options are:

1. *Modular design*. Create modules that can be separately fabricated, assembled, and tested. Make them correspond to different options that fit together interchangeably for final assembly and test. Option maintenance can then be reduced to replacement of modules. Developing a modular design structure is not so hard during original design. Retrofitting into it is tough.
2. *Sequential buildup*. Try to design options that are simple, step-by-step buildups. Production flow does not have to start and stop or double back on itself. The idea is to be able to fabricate, assemble, and test streams of options for different customers without breaking rhythm. If fabrication of option parts is only substitution of tooling in a fast setup on the same equipment, that simplifies the situation.

If options require a unique buildup for each different combination, the production situation is little different than if each order were custom designed. The objective of options is to provide the variety without this expense. Suppose a particular option requires special holes in the frame of a machine. When the option is to be fitted, access to drill the holes may be impossible without partial disassembly, which adds expense and potential for damage. Better to drill such holes when the frame is started; better yet to drill all holes needed for any option when any frame is started. Best is to add any option without any special holes for it alone.

Structuring options is very important for company communication. An order for an option should be unambiguous and therefore translatable into every activity necessary to provide the option—including pre-drilling special holes in a frame if needed.

Option coding forms the basis for sales literature, order entry systems, materials systems, operation instructions, and cost systems. If either the engineering or systems coding of options is not well done, the company carries a huge burden and cannot avoid waste in serving the customer. Excess communication is waste. Errors in communication are a quality problem.

Some methods for coding modular bills of material and master build lists are ingenious, but complex options create problems. These are usually described with some mind-boggling number of combinatorial possibilities for all the options on one base-case product. Consequently, the production department suggests discontinuing unusual combinations, a solution vetoed by those who wish to please a broad market spectrum. No known definitive studies on the waste in dealing with options have been done; however, according to companies that have them, the problems are:

1. Options require more parts in total. More parts are harder to manage than fewer, and they complicate efforts to automate.
2. Low-usage option parts are hard to manage when mixed with a high-volume product. Their visibility is submerged, so getting the right part to the right place at the right time is hard. Forecasting them is more difficult.
3. Options give people choices and opportunities to change their mind, and they do. It is easy to spend much time defining an order and then checking "if and when" optional use material can be obtained—a customer-driven but material-constrained business.
4. The pricing is complicated. Some options are high-margin while others are "toss-ins." Much selling effort can be ex-

plaining options or determining if what the customer wants is really suitable for their use. It complicates the customer's choice.

5. Problems with bills of material and option-coding systems seem to be:
   a. Interpreting conditions of use for parts and options—communication of Boolian-type logic statements, which are combinations of "and-with-or-if and not."
   b. Bills of material must be reflected in several different forms in different databases. Coordinating changes accurately is a maintenance problem. (For an example of a large number of parts, multiply referenced with different codings and numbering systems, refer to the long rack of catalogs across the counter of the neighborhood auto parts store. What would happen if making an *accurate,* coordinated change in all of them became necessary?)

Despite the problems, customers want options; thus the question is how to attract a customer base with options structured to produce and service them with little waste. There seems to be no one-answer-fits-all solution. It depends on the type of product and the purpose of the options.

For some, perhaps, offering a base case loaded with newly standardized options makes sense. (Then some customers will want stripped versions as options.) Some options (especially trim and hang-on) may be added by a dealer or a service department.

From an operations viewpoint, the best solution results in the simplest bills of material at all points in the total process. Most companies with high option content want positive bills of material; that is, they want people and systems at all points in the process not to interpret Boolian logic statements of use, substitution possibilities, or condition-of-use restrictions. Just build what it says—straight out.

Doing this starts by simplifying how options are designed and built. Complex costing and pricing of options at best creates a value-added cover for the issues for as long as the cover lasts. It does not address providing the product and service needed by the customer without waste. The way options are structured reflects how a company works, which cannot be changed overnight. However, as the company improves its capability, the simplification of option structure will become both an opportunity and a necessity.

COMPUTER-CENTERED SYSTEMS

The technical vision of the factory of the future is a plant operated as one big, computer-controlled cell. A part can be computer designed (computer-aided design) and its fabrication instructions can be generated by computer-aided manufacturing (CAD/CAM). Then the correct material can be automatically selected and automatically moved through the computer-controlled equipment necessary to fabricate and assemble it. Along the way, the part will be subjected to computer-generated test procedures. Link it all together and you have computer-integrated manufacturing (CIM).

Limited instances of CIM are working today, but a major software problem is linking all the various systems. In most companies, islands of automation operate independently of each other because the computer languages used by each cannot traverse the gulf between them, but headway is being made on that problem also. General Motors' manufacturing automation protocol is long step toward developing a standard software interface between machines and systems.

Artificial intelligence (AI) is a phrase denoting software that replicates features of reasoning; that is, it uses a set of decision rules to make limited judgments that go beyond just applying straightforward formulas to data. The best known "laboratory" versions are programs that play chess—and win against all but the most expert human opponents.

In many companies, a materials system and an accounting system using computers is old hat. Order entry is by computer. Distribution inventories are controlled by computer. Sometimes all these systems are rolled together into one big system with exchange of data. Material requirements planning is extended to link with capacity, accounting, financial, and logistics data, perhaps into one grand approach called manufacturing resource planning.

The potential in such approaches is great if the people using them can surmount the problems. Computer systems do not correct basic problems by themselves. People do, and more can be wasted with computers than without them.

A technical discussion of CIM is beyond reach here. The key point is recognition that computerization of manufacturing provides technical possibilities to create waste as well as eliminate it. Managements sometimes incorrectly assume that automation is synonymous with manufacturing excellence. They are too much fascinated with the technology and too little fascinated with the totality of human development necessary to make it work well.

A computer is a tool. AI is a tool. A robot is a tool. Linked together, automation still consists of tools—to be used wisely or unwisely. If the name *robot* did not suggest humanoid characteristics, robots might be better applied as tools.

In the end, *people* make anything work. Companies striving to make CIM work tell the same story. It takes leadership from the top and an integrated effort by all functions of the organization. People must learn to think in a new way. Computer-integrated manufacturing is really people-integrated manufacturing. People and process development come first—pre-automation preparation.

Computerization alone does not identify waste, correct many quality problems, rethink how to structure the design of a family of products, set a marketing strategy, or make suggestions for improvement. It does create need for disciplined integration such that these problems must be addressed if they can be seen through all the systems surrounding them. If basic disciplines can be achieved in a simple way, the expense of automating the waste can be avoided. Automate where waste is eliminated by doing so.

CHAPTER SIX

Attaining Total People Involvement

Once sure of future direction, the most critical factor of success is development of people. As soon as they are ready, give the action point people responsibility and expect performance. Even if skill development has not progressed much, the change in work atmosphere produces improvement.

A Navistar foundry improved quality by giving workers basic instruction on quality and the responsibility to do the job right. Workers at the final operation grind castings, inspect them, load them for shipment, and attach a quality certification tag, which they sign personally. The tag identifies the part and its source. It proclaims quality to the user and says that, in case of trouble, telephone the *worker* who signed the tag. Next to the grinders is a telephone—the *workers'* telephone.

Now the workers know by telephone voice some of the people in far-off places who use their castings. Their interest in quality perked up. They do not hesitate to inform other workers in core room, furnace, or pouring when something is not right. Some of the "bureaucracy" of staff and supervisors between customers and workers is gone.

Call it job enrichment, job enlargement, or what you will, the added responsibility did not increase the work cycle times of the direct workers and it was real participation, not a make-them-feel-like-they-are-participating sham. The improved quality made their job easier, not harder, and it increased the productivity of the "bureaucracy" also, provided they understood.

Management and staff acceptance of true worker responsibility comes hard. Most workers welcome the approach. It is better than being treated as a child, but they will naturally question the sincerity of management, and management doubts the capabilities of workers. Some workers may only be capable of limited development; some are too immature to accept responsibility; but in general, workers accept this change easier than management and staff do.

The job of management and staff is to enable the direct-action people to perform better. That job is coaching, and coaching is development of others while developing yourself. Coaches teach, make correction, blend talent as a team, develop individual and collective skill, and motivate, but coaches do not play themselves. That is reserved for the players.

Even planning and coordinating jobs should be done with intent to enable improvement and much of that ultimately by action-stage people. This is a matter of basic attitude. The objective of overhead functions in a business is to support the direct-action work. The objective of line work is not to support the overhead, a deception easily inferred from the way costs are assembled.

CLOSE TO THE ACTION

Staff distant from the action grow to accept their daily rounds of reports and crisis chasing as the way work should be done. It all seems energetic and spirited, and so it all must be vitally important. Amid an ocean of waste, a vision of a world consisting of anything else is hard to come by.

Staying close to the action cannot be done with many people and little action. A lean staff is a busy staff. Lore has it that the nine-to-fiver will outperform the 60-hour-per-week staffer because the workaholic is probably dithering and disorganized. That assumption is dangerous. Suppose much of that time is spent figuring out how to make workers ever more effective—"fighting a war," not twiddling a system.

Many hours on the job make communication easier. On the scene means in touch—with action people and with others. Meetings are not put off because of vacations, nor is time spent recapping what happened for those not present earlier. Telephone tag is reduced. Being there is much of the battle. In sportscasting parlance, it is called "presence."

In JIT/TQ plants, managers and staff should visit the operations regularly. If the plant floor is developed with workplace organization, a staff person will see a lot when passing through. A plant manager will want to see more. Problems are understood firsthand rather than filtered through reports—and those take someone's time to prepare.

A side benefit is achieved by expecting all managers to be in regular contact with action areas. Too many managers are obvious. Problem visibility applies to more than just shop floor operations.

Regular meetings promote easy communication. A plant manager recently decided to hold an operations meeting every morning, just after the plant manager's tour of the floor and before telephone prime time for everyone. Fear was that time would be wasted if no immediate problems were pending. As it turned out, many meetings were short but valuable. Everyone received the same signal from the daily huddle, and it was a good place for little side conversations that prevented the need for other meetings and phone calls.

This management style fits the company warring on waste. If inventory is cut, the response time to correct problems is cut. Managing ongoing operations becomes more like managing an air traffic control tower. However, if the approach to management does not change, the reduction in inventory brings little benefit because *the reform of management itself is the real goal.*

Incorporating automation does nothing to change this basic management approach. Those making the automation run are the action people, even if they rarely touch the product itself. A reason often given for automation from the "engineering mentality" is to rid the company of direct-labor problems. If by that they mean that there is no longer a place for someone semi-literate, there is some truth to it. If they believe that people problems will go away, they are wrong.

People problems depend on the kind of people brought into the company and, most of all, on the approach of management. One of the most difficult union-management problems in recent American history has been the air traffic controllers—skilled work in a high-tech environment. Working for constant improvement takes close association with the action people no matter how technical the work.

A company with enough cash may temporarily escape a bad labor relations history by moving a plant to a new location, either on-shore or off-shore, but it cannot escape itself. In an industry where continuous improvement is the norm, the human side of managing must be mastered no matter where the action takes place, or by whom. Changing us is not as swift as changing location.

RESPONSIBILITY AT THE SOURCE IS LINE MANAGEMENT LEADERSHIP

Manufacturing excellence requires line management leadership. The effort cannot be integrated without it. Responsibility at the point of action demands it.

Strong staffs have rivalries, and rivalries prevent an integrated point of view. Staff groups can create enthusiasm for total quality or just-in-time programs of one kind or another, but they cannot reform the organization.

If a staff group is charged with implementing something called JIT, the concept is seen by others in narrow terms. If pushed by purchasing, it must only be a program for suppliers; if by production control, it is only a different way to manage inventory; if by manufacturing engineering, it must be an equipment project; and so on. The staff members may have a broad concept of the total effort, but they cannot bring all the pieces together.

The result is a common phenomenon: "pockets of excellence." A portion of a plant shows great improvement. Costs and quality improve on part of a product line, but the effort is confined because the surrounding environment cannot support it. The pocket itself is in danger of collapse if the middle management champion is displaced. Momentum to continue a long-term, integrative change can only come from sustained line leadership at a high enough level to change the structure of organizations and the way people think.

The more successful production area changes have had a common beginning. A group of people, including both line and staff, underwent a gestation period of some months while learning what to do. Then they went into the shop and started. A project leader might have coordinated, but the production managers and supervisors took a lead role. Without it, shop-level changes are stymied.

If no more than this is done, "techniques are installed." To get into the spirit of workplace organization, worker training and involvement must begin, else basic problem solving is not sufficiently transferred to the workers and supervisors themselves. That is responsibility at the source in production. To achieve it, the workers must increase both in knowledge and responsibility, and the supervisor's role evolves to facilitating.

A supervisor used to bossing around in detail cannot accommodate to this easily, so supervisor preparation must preceed significant changes in the roles of workers. No supervisor wants to be thrust into something they do not understand mentally, and the "bossy" ones will have trouble emotionally.

The basic nature of the problem is well known—the age-old management problem of accepting delegation of authority and responsibility. Line managers still exercise discipline but change the way they do it. Workers with responsibility must perform. A common fallacy is that participative management is permissive management. Not so.

Workers performing as a team tend to exercise a certain amount of peer pressure, which helps discipline. Someone who must always have a fellow worker cover them will probably hear about it from peers, and that strategy works for many. Flagrant nonperformance has to be dealt with by the supervisor, however, and failure to do so demoralizes the rest of a work group who are really trying. Manufacturing excellence and participatory management are no cure for every shortcoming of people at work. Alcoholism, divorce, and disinterest for other reasons do not go away.

This change is easiest for production supervisors to understand as maturing into delegation of responsibility—a higher-level management approach. In any case, the transition is tough, an ego-breaking experience for some. Not all are able to make the change, and a barrier to progress in several companies has been the time required to develop supervisors with the quality to handle it. A compensating factor is that fewer full-time supervisors may be required. If production firefighting decreases, a major time consumption for supervisors goes away. They have more time to pay attention to people problems—and to work out how to do the job better. A full-time supervisor can be augmented by several lead workers for large groups.

The initial changes usually begin in production, and the most successful have been characterized by strong leadership and action. Less successful ones are characterized by much training and planning but little doing. Both the training and the action are important. The best training is followed very quickly by doing because people learn best by doing. (Japanese companies going this route refer to "dirty instruction": hands-on, show-how sessions first, followed by doing, and finally by explanations of why the approaches work. This method seems to work well in other cultures, too.)

Integration of Effort

Management excellence cannot come from fragmented contributions by various functional staffs; that is, if quality assurance has exclusive jurisdiction over a quality program, if production control has an inventory program, and so on. Each staff seeks to impose another set of techniques, each set demanding adjustments and attention from an already choked line organization. Only so much can be assimilated at a time, and it should be cohesive.

To keep from choking, line management must maintain consistent direction. Staffs need to be integrative in planning, which is no small demand for those steeped in their own specialties and which implies much about the personnel management of staff groups. No human can be omniscient, but most can acquire a perspective from different points of view, and by asking them to work directly with workers and line people, they also stay in touch with reality.

As observed earlier, design engineers have a highly integrative role unrecognized by many. Though stereotyped, they are perhaps no worse than other staff specialists. Perspective may broaden from training programs but, practically speaking, experience with other functions in their own company is better. If they have spent time with real customers and also in plants building what they design, their concepts expand. Even if they are, say, the world experts in ultrasonic effect on crystallization, designers need to know how to apply their knowledge to something buildable and marketable.

The argument for cloistering specialists is that concentration builds their expertise. However, an old stream of research on scientists and engineers indicates that researchers are more productive when they have a diversity of interests.[1] A little broadening may prevent them from going flat and certainly assists in creating something that needs less rework by the rest of the organization. Perhaps the same can be said of other staff specialists.

Japanese are good integrators. One reason is wholistic thinking, born of Oriental religions and expressed in management as a con-

---

[1]Pelz and Andrews, *Scientists in Organizations* (New York: John Wiley & Sons, 1966).

sciousness of how one's immediate activity relates to the whole. Meetings often start with the obligatory recitation of overall objectives (to put minds on the same track) then plod through a diagram—incredibly detailed by American standards—showing how the detail fits into a whole fabric. Since the human mind is finite, this thinking process simplifies much detail to make it fit together. If left undone, organizational detail sprawls in all directions.

Americans are accustomed to thinking that integration of a company is only in one place—the top, where strategy is made. Well-managed strategy is integrated in many places. The sales representative meeting the customer brings along everything a company represents. The operator assembling and testing a product puts together the physical representation of another mass of company detail. A product prototype brings together everything engineering can assimilate.

Integration in strategic terms is not sufficient. Execution of the strategy needs to be integrated in detail.

Integration takes place by keeping focused on an overall goal and by promoting policies that keep the detail meshed together. Since continuous improvement on a broad front can lead to discontinuous activity in many different areas, a stepwise progression through several primary goals is useful. All depends on the specific company case, but a typical goal progression is:

1. *Quality improvement.* All of the "techniques" from workplace organization to preventive maintenance may be directed toward this in the beginning. Elegance in meeting other objectives is hindered until substantial progress is made in meeting this one. (Prompt delivery of junk is a poor goal.)
2. *Dependability.* Doing what was promised when promised. Prompt delivery of top quality adds value.
3. *Cost reduction.* Do it for less. Great strategy if you can already provide quality fast.
4. *Flexibility.* Put some variety into it. Roll with the changes.

Patterns of Thinking

A popular business witticism is that impending bankruptcy, like a gun pointed to the head, concentrates the mind beautifully, but on what is left unstated. In an emergency, managers do not think much differently from the way they always have, only more quickly.

Managers start emergency cuts with travel, training, or whatever else will not affect revenue in the short run. Seldom is there time for a thoroughgoing operations review because that, to be effective, has to be a part of the daily pattern of thinking.

The pattern of thinking must piece puzzles together in simple ways. Cost models of business situations do not do this adequately. If the thinking pattern is primarily in terms of budgets and accounts, much of the importance of quality, timing, and flexibility is lost. If one thinks only in terms of customers and sales, much of the importance of operations costs, delays, and timing is lost. And if one thinks only of production, the importance of the customer is lost.

The gist of many business school case studies is the mess created by this divergence of thinking. The emotions create an exciting script. The most common student solutions are to reestablish and affirm a consistent set of business goals. That the various protagonists will either subscribe to them or be fired is taken for granted. The necessity of much work at low levels, making sure everything dovetails, is seldom recognized. However, if one reads the accounts of companies deemed excellent, that is something they do well, and on which they expend a major fraction of organizational energy.

Staff as Coaches: Tough Role Change

One major problem of JIT implementation, in Japan as everywhere, is the reluctance of professional staff to be advisors and to work close to the action. Many have great difficulty comprehending that when it becomes personal. More prestige seems to accrue to working with an advanced project having a big budget than to assisting with the details of a series of simple improvement ideas.

Attack staff work that tollgates the flow of production or the flow of information, especially customer orders. Promote staff work that breaks such jams.

Tollgating stops a flow of traffic to collect a fee. Great tollgating empires have been created by eking an existence from the work of others—until the source of traffic through the tollgate disappears. Much staff work seems to require stopping other peoples' work to perform some check or add a specialized contribution. Production control stops material to count it. Accounting stops work to cost it. Quality control stops work to check it. And legalists must stop work to be sure procedures are observed and rights protected. All are honorable missions, but could they be performed without tollgating anything? Could it be done more by coaching the action people? Perhaps, but tollgating automatically defers to specialist dignity. Coaching must provide dignity, too.

Staff work is not diminished in importance if tollgating functions come under close scrutiny, but functional specialists may not see it that way. Making the production worker responsible for quality eliminates quality inspectors. Having gauges and instruments close to the workers reduces business in special inspection departments. Such moves pop the bubble of big quality control organizations. Likewise, an inventory manager is apt to feel a certain unease if a declared objective is to eliminate stored inventory.

The role of the quality staff shifts more to teacher and coach. If line workers are responsible for quality, they must know how to make measurements and how to avoid defects. The need for expertise from quality specialists is not diminished. It really increases. Some measurement problems at most manufacturing companies require special attention, thus all inspection cannot be absorbed by the workers, and keeping the gauges and instruments working to a uniform standard is always a problem. The role change is that the quality specialist has to work with the worker and work to promote production flow, not tollgate it. If the production worker needs coaching, then so does the sales representative, but perhaps not in the same way. Engineering and production may put a quality product into the field, but most customers perceive a company's quality through the actions of sales and service representatives. Service quality done right the first time is the responsibility of those who do the work.

By contrast, the need for manufacturing engineers to unplug work flow may call for greater staffing in that area. Companies like Omark Industries, who has made a good start improving manufacturing, all say that if starting over, they would strengthen manufacturing engineering more quickly. They are not gearing up for major automation but rather working through the myriad tool redesigns and machine alterations necessary to eliminate defects, reduce setup times, and change layouts.

Omark strengthened manufacturing engineering by eliminating wasteful paperwork so that qualified people could shift more attention to the technical problems—the real engineering that engineers like to do. New engineers were hired only at small plants that had none.

However, manufacturing engineering is ineffective if engineering work cannot be executed, and that requires capability in tool shops, maintenance shops, and other places where gadgetry is built. *A plant that cannot perform a large percentage of its own tooling and machine modification in-house is severely handicapped in continuous improvement.* Trials of devices are slowed if they must be shipped in and out for revisions between times. Most important, skill development greatly depends on thoroughly understanding tooling and equipment in the context of their intended application.

Few plants can be technically self-sufficient in everything, but their personnel should be expert in the heart of their own technology. Farming out key tooling work to save cost in the short run may increase cost in the long run. Scattering the technical know-how among diverse tooling and equipment suppliers is strategically questionable.

The work of manufacturing engineers consists of many small grit-and-grease projects that demand expertise beyond operators and technicians. A busy manufacturing engineer is likely to put off such projects while working on something that seems more important, like developing a simulation of the process if a terminal is handy. "Working with the worker" means giving priority to the details of improving the current process as well as future processes.

This role change is difficult for some. In one plant, the work force enthusiastically endorsed the JIT/TQ process, but they revolted against the engineers (in this case, quality and manufacturing engineers) who would not help them overcome the problems they uncovered. They taunted them and hung derogatory signs. The workers' action was unusual, but the situation was not. Engineers who will not work close to the action cannot do the job.

Methods improvement is another discipline in which coaching the work force is important. They need basic instruction in motion economy, layout, work methods description, time study, man-machine charting—time-honored basics dating back to early in the century. Today, plants often have no one to coach these skills, or the industrial engineers have limited experience in them.

Many schools of industrial engineering have skimped on this material in recent years, favoring more advanced technology and model building. Production managers have cut methods engineers during budget freezes on grounds that methods improvement is so elementary anyone could do it. They are right, but exposure to the basics in practice is still needed. Perhaps lack of association with the basics is why engineers concentrating on more advanced technology will program a robot to perform work that is either poorly designed or totally unnecessary.

Reforming the staff work that less directly supports production requires digging deeply into the real reasons for it, some of which take no end of twists and turns. Customer orders are delayed to interpret whether any significant liabilities lurk within. Returns from customers are delayed so they can be counted in a future period's warranty expense. Layout changes, in production and elsewhere and no matter how trivial, must be checked by a certified safety officer lest the insurance premium become extortionate. The source of these delays has been expedient responses to knotty predicaments, which over time accumulate to a thicket of tollgates.

Simplifying this thicket entails sorting through a labyrinth of complexity to separate the necessary from the waste. As a guide to thinking, go back to Chapter 2 and review the seven wastes. They apply just as much to nonproduction staff work as to the shop floor, and many situations are analogous to long setup times and so forth.

PERFORMANCE MEASUREMENT

Performance measures historically used for production are seldom in accord with the objectives and concepts of continuous improvement. Performance measures reflect the values and expectations of those who desire the performance—not always consistently, but their pattern of thinking is evident in them.

Performance measures for production derive from performance measures used for an entire company, depending how managers think production relates to the whole. Historically, companies emphasized return on investment as a prime criterion of success, looking at it from an investor's viewpoint, and frequently they emphasized growth of sales and assets. Beyond that, the measures emphasized market leadership, being a good corporate citizen, and similar factors lauded in corporate annual reports.

When a company shifts its pattern of thinking to continuous improvement, the change in goals mostly reflects *how* it will earn more money (or just survive) over the long haul. The annual reports speak of quality, dependability, customer service, productivity, cost reduction, and, very noticeably, human relations. (Omark issued its 1984 annual report with the names of all employees on it—over 4,000 names.)

This change in corporate perception of what it is about is profound, but that will be addressed in Chapter 10. While all companies must take in more than they spend, this changed outlook strongly affects the performance measures used for production, and the change process is another block-by-block construction job.

Performance measures affect careers and promises made to people, so they are not changed without turmoil. Managers who have learned how to cope with a system they know well are thrust into a new arena at an advanced career stage. Changing everything at once is shock treatment.

Purging the Old

Begin by deemphasizing performance measures that distract managers from long-term improvement. Two of the most common are equipment utilization and labor efficiency measurements. Production managers driven hard by weekly reports of such measures tend not think far beyond them. Their compromises lean in the direction expressed by a supervisor in electronics assembly, who summarized his life thus: "Keep the line running, argue every case with QC, and keep 'em out of the john."

If efficiency measurements make managers nearsighted, incentive payment systems are near-blinding. Individual incentive systems make a shop into an every-worker-for-himself affair, which is disastrous if they must work as a team in flow-type production. Where people must work together, group incentives are used. Those are almost as bad. They still tilt all decisions toward the next paycheck.

At worst, incentive systems penalize the worker's pay for an improvement in production that is recognized as a change in method. Then, the standard for incentive purposes is tightened, so pay is effectively decreased. The workers' incentive is to improve production by every means right up to the point where a new standard might be established, then stop.

A number of JIT/TQ reforms have been attempted (with partial success) in incentive shops. Most managements believe they will find a way to reconcile the ideas. None succeed, and all sooner or later must go through a donnybrook with the work force if they wish to convert to a straight wage plan. The high earners are incensed by the pay cut; low earners get an increase. The usual approach is to try converting to a straight-rate system with a small increase in average pay, thus as few people as possible take pay cuts.

Incentive systems undermine efforts to cultivate a flexible work force. People are reluctant to move to jobs that pay less, and even if this can be administratively worked out, it is complicated. Incentive systems go along with narrow job descriptions, administrative burden, bidding rights on jobs, and informal pecking orders for "gravy positions."

Actually, many measurements used for production make a very bad assumption: the design and improvement of production is a staff concern, and line management and workers are only to operate what has been handed to them. All performance measurements need to be examined to determine the extent to which this assumption pervades them and then changed according to the notion that staff, line management, and work force are all partners in constantly improving it.

Costing It Out

Budget and cost structures can also drive continuous improvement to destruction. A move that reduces waste can appear to increase cost and vice versa. A typical example: At John Deere's Horicon Works, assembly operators had trouble keeping in stock the proper diameter hoses cut to the right length. Hose was purchased in coils, which were machine cut in lot quantities, boxed, and taken to assembly. The solution was to keep a coil of each hose diameter at assembly and to let assembly operators cut it to length with a knife.

However, labor time for a machine cut was less than for a knife cut, so the cost appeared to increase. Buried in the overhead was the cost of several different containers, inventory investment, the indirect labor of 17 different "moves" of a hose (nine times touching it plus eight times moving it) on its way through the plant, and the planning necessary to make it work. (Deere made the change.)

Production cost structures often apply high overhead percentages to direct labor. A wily manager soon learns that an easy way to fake a cost reduction is by reclassifying direct labor work to indirect. Sharp cost accountants are well aware of such possibilities, and experienced auditors can recount a full repertoire of games that can be played with cost models imperfectly portraying reality.

The best cost systems cannot accurately portray the details of reality, but official sanction for production methods changes is to pass a cost savings test. A typical quandry is that throwing away defects seems to be less costly than any corrective action proposed. The cost system does not detail the delays and other wastes of defects. They are buried in the allowances and overhead. If a "petty cash" alteration obviously reduces waste but cost figures show nothing worth going for, make the change anyway.

At the same time, beware of spending big money without thorough examination. A series of little improvements can add up to something that might not have been achieved by an expensive change in tooling and equipment anyway. In pursuing the source of quality problems, one finds that they may not be where they were thought to be. Likewise, the real sources of high cost may not be where the system assigns them.

Allocate improvement money to line managers and judge results by whether costs aggregated by plant or by department trend downward. Aggregated costs pool together all the direct, indirect, overhead, and miscellaneous. If real progress is made, those should decrease. If not, line management leadership and judgment must be reviewed.

Coming Clean

Methodologies for devising many of these performance measurements are represented by books in a library of business. Changing any one of them would create ripples in a placid management pond. Revising the philosophy by which a company does business drains the pond completely. Companies find that all the methods and systems put in place over many years have to be rethought because they rest on assumptions about manufacturing—and about people—by which they no longer want to live. Go gack to where it is simple again and start over.

For an old company, the stumbling block to manufacturing excellence is that a comfortable executive does not want to take on a complete purge of an old management culture. Such matters as setup time reduction and cell layouts are nonthreatening. Rebuilding the thought pattern of an old organization tests the mettle of the most ardent evangelical.

Metering in the New

The long list of performance measures in Table 6–1 does not imply that performance measurement of continuous improvement must be complex. Excessive measuring is waste also, and the objective is to improve performance, not measure it to death.

Some of the JIT/TQ pilot projects whose founders "just went into the shop and started" found after a few months that they were not documenting their achievement. An overcorrection gathers so much data that progress bogs down in its reporting. Most managements must fumble at first when developing reasonable goals and performance measures for themselves.

An old organization cannot accept all these measures at once. They must be adapted to the context of the company and metered in as the organization is ready for them.

A few measures should be used to set targets for improvement. Others are merely monitored and go along for the ride. The ones used for goal setting at the top of the organization should be few in number because too many different targets at once jumbles the pattern of thinking and sets people working at cross-purposes.

Back in Chapter 3 the general manager of Figure 3–4 set one primary quality goal: Reduce emergency field service calls by half in one year. That set off a flurry of work in field service, marketing, engineering, and production. Each was able to establish some subgoals, which should contribute to the overall one, then monitor their performance by it. The best production could do is be sure the product is 100 percent when it leaves the plant and to revise packing and shipping to better assure that the product arrives in the same condition.

Production could come close to its goal by inspecting and reworking the product to death, but that is costly. Better to set major subgoals: Decrease failure rates at final functional test by half in six months, and decrease failure rates in burn-in (reliability) testing by half in six months. To accomplish those goals, production must work with design engineering as well as make improvements in its

TABLE 6–1 Performance Measures for Production Organizations

1. Recent *trend* in primary quality measures (defect rates).*
2. Recent *trend* in work-in-process inventory (prime indicator of waste).*
3. Long-term trend in cost of goods sold (or other aggregate cost measures).
4. Linear production (repetitive case) or production to schedule (job shop case).
5. Customer service level (from stock) or customer lead time (to order).
6. Skill improvement of people (managers, staff, workers).
   a. In depth (quality checks, setups, interpreting orders, etc.).
   b. In breadth (cross-training for different positions).
7. Condition of plant, equipment, and people (Many of the following must be seen in person to be judged).
   a. Total productivity.
   b. Morale.
   c. Workplace organization.
   d. Visibility of conditions.†
   e. Equipment availability and maintenance status.
   f. Layout (material travel distances).
   g. Absenteeism.
   h. Safety.
   i. Suggestion rates.
8. Capital budgets.
9. Departmental expenses.
10. Other measures, some of which are vital to some kinds of manufacturing missions:
    a. Response time to execute marketing mix changes.
    b. Response time to execute engineering changes.
    c. Introduction time for new products.

*All measures must be compared with sales or production levels to be placed in context, and all managements must monitor expense to be sure it is not flying away. That is basic to any business, but after that, if management wishes to monitor only a few numbers, these are two key measures. If a production organization is performing well by these, honestly and consistently taken, it is hard for it to show poorly by any other measure.

†To check how well it is *really* going, look for the evidence of problems being open and easy to see, both on the shop floor and in staff and support areas.

own processes; thus the direction of engineering improvement and production improvement must be complementary. Unless the departments work together in detail, their improvement efforts will clash.

No one knows whether these targets are reasonable, not even the manager who sets them. If they motivate, they are reasonable. The important goal is stimulating many parts of the organization to take actions that should all mesh together in the same direction.

Each group achieving its target just by its own improvements is not sufficient. Engineering's goal of 300 percent increase in mean-time-before-failure might be achieved on newly released product through improved design alone. On its own, engineering may "harden" design entirely by adding expense—tougher specifications and higher-grade components. Another way is to work with production.

For production submanagers to set contributing targets, they need to understand how their action assists the overall effort. The initial reason for setup time reduction, for instance, is to improve quality, and the improved material flow contributes to that; but the idea is not to sacrifice quality just to attain setup time reduction. The increased setups per month are monitored but, for a time, they "just go along for the ride."

As improvement makes progress, general management's primary target may shift to something like decreasing the throughput time of material while holding quality levels already attained. Then, the purpose of reducing of setup times would shift. The set of performance measures might remain nearly the same, but the emphasis on various ones would change.

The purpose of performance measures is to make the company better—always better. This reasoning underlies the construction of Table 6–1, which is a list of performance measures that may vary in emphasis from company to company and from time to time. A framework of targets and performance measures means nothing unless managers and workers understand how they relate to them—and how the framework of production on which they are based goes together. Use of these measures by caretaker management accomplishes little.

Since all the measures in Table 6–1 are not self-evident, commentary is provided on selected ones.

1. *Quality trends*. These do not go down and stay down. Even the best companies must reestablish quality after a change in products or an influx of new workers. The importance of trend is to monitor whether improvement is always taking place, given the recent situation.

2. *Work-in-process inventory*. This is a major indicator of management status in a production facility, but its existence is not wholly due to factors within the facility's control. WIP also cannot

always stay at minimal levels; it increases due to new product, new workers, and similar situations. The trend should show that progress is always being made: If WIP goes up, it starts to come back down again.

3. *Cost of goods sold.* This measure responds very slowly to continuous improvement efforts because it takes time for people development, tooling development, improved maintenance, and so on to show results (like waiting for the teenager to grow up). After two or three years, the trend should go downward if all other things remain equal—which they probably will not.

4. *Linear production.* Largely the subject of Chapter 7, this refers to leveling and balancing the short-term load in production—attaining an equal amount of production each day. (In many plants, production moves at a snail's pace, particularly in final assembly—until month's end, when it roars into overtime to make schedule and month-end billing times. The phenomenon is caused by large lot sizes and the entire approach of manufacturing management.)

5. *Customer service.* Production improvements that do not improve on this do not accomplish much. This too is greatly affected by both scheduling and the total approach to manufacturing management.

6. *Skill improvement of people.* Perhaps this should rank number one. Productivity improves *only* because, in some way, the same number of people can accomplish more. If managers are to put major effort into development of those working for them, their performance in this regard should be measured.

7a. *Total productivity.* Productivity is improved by small gains (such as reduction in rework). If productivity is measured, it should be overall productivity:

$$\frac{\text{Measure of output}}{\text{Total headcount}}$$

Concentrating on direct labor productivity misses too much. Include all the indirect, staff, managers, and part-time window washers too.

7b. *Morale.* This intangible makes up for many deficiencies elsewhere, but it should be part of continuous improvement—a big part. Outstanding performance on many other measures cannot be attained without high morale. No matter how automated production is, nothing works unless the people involved want it to.

7c and *d*. *Workplace organization and visibility*. These are tangible but can only be rated, not given a direct number. These measures are two vital signs of morale as well as stimulators of other action.

8. *Capital budgets (for plants)*. All major capital expenditures must be approved by management. The issue is capital for the small improvements that allow layout changes, setup time reduction, and driving defects to near zero. These activities are hamstrung if a committee has to approve every $500 appropriation. Allocate such capital to department managers, and expect them to use it wisely but stay in budget.

9. *Departmental expenses*. Variance reports detailed by operation seldom suggest anything that could not be seen otherwise, and accounting systems applied too finely have many flaws. Nonetheless, no company can spend money it does not have, and controllers are charged with seeing that they do not. Keep the accounting for expenditures simple but strict. Expenses aggregated by department seems a logical low level for budget accountability (not the same as detailed targets to reduce wastes).

WORK FORCE PARTICIPATION IS MANAGEMENT PARTICIPATION

Genuine work force participation with management largely comes from management participation with the work force. Regardless of the ugliness of past union-management relations, hope exists if a manager can talk directly to most workers in detail about what they do with their hands and why. Confidence, attitude, and trust make the difference.

Selection

Start with selection of people—not workers first, but managers, staff, and supervisors. If a new plant is starting, those come first If not, some form of evolutionary reselection process will be needed. Hopefully, some of the unfit will select themselves off the payroll.

What to look for? After resumes, predictive tests, and attestations of prior performance by former associates, the interviewing process starts to zero in on attitude. Attitude is crucial in two areas:

willingness to be on the job and stay there and ability to work with other people.

These attributes are hard to evaluate because they only show up over time. They do not show up on tests, and a good actor can cover them. Best advice is do not be in a hurry to hire and do maintain a long probationary period. Some fast-trackers with great ability may be lost that way, but the final team should be more stable, which is very important in sustaining effort.

Golden chain techniques are well-known methods to increase the chances that people will stay with a company, but nothing is certain in the United States. One can only provide a challenging environment in which people will want to work, and strive for the best retention average possible. The chances are improved by avoiding job-hoppers in a rush to hire.

Even when selecting lead managers for JIT/TQ (by whatever name), willingness to stay with the company is important. Other companies excited by the prospect of an experienced manager heading their own JIT/TQ will lure them, and consulting also offers the promise of comfortable money for travel experience. Many fledgling efforts have been crippled this way.

One can ask key managers to sign longevity agreements as a signal of intent or whatever, but outright slavery is passé, so there is no keeping one who wants to leave. If other companies raid personnel, at least they recognize the company for doing something well.

Less excusable are personnel transfers of lead people early in JIT/TQ conversion. A few quick whacks at inventory, space, and defects do not permanently implant the approach in the organization. Developing people takes a long time, so all personnel moves should have in mind the long-term development of *everybody* involved (something like planning a church's missionary program).

Evaluating a manager or staff specialist for ability to work with people is also better done long-term than short. Managers and staff should have the ability—and willingness—to interact with anyone: top executives, workers, technicians, customers—anyone. Good people skills coupled with technical talent are in short supply, so perhaps they can be developed if people with any inclination for it are hired in the first place, and if they understand *why* they need to work with people on the shop floor, in tool shops, or anywhere else.

Abilities to communicate and to work with people are attributes executives mention most often when asked what is important in making a good manager at any level. JIT/TQ intensifies the importance of these managerial characteristics.

As for workers, a company does not have to hire whatever antisocial flotsam first fills out an application. Workers should also have interest in working with other people. Equal employment laws in hiring and promoting do exist, but the criterion of communicating well can be clearly stated so it is not an unnecessary discriminator. Ask prospective employees to communicate some simple instructions to fellow interviewees, for example. Ask them to perform some task with a group, and look for clues to later behavior. If they will eventually be asked to describe and improve their work as part of the job, demonstration of an interest in doing that is not an unreasonable request. That job requirement should be documented for employment screening.

Large manufacturers are experienced in such things. Small ones frequently make hiring errors. With no personnel specialist available, they quick-sort applicants when need for help is desperate—if applicants can be found at all. That situation is the beginning of difficulties in developing people.

Job Security

The root of many problems is fear of unemployment. A company intending to increase productivity 20 to 30 to 50 percent has a serious credibility problem obtaining participation from a work force who believes that their reward will be an early place in the unemployment line. If costs drop quickly, perhaps the company could take market share from competitors and ship their unemployment to them, but that has not happened in the pioneer American cases. After major productivity improvements, companies still fight for life against tough competitors.

Suspicion of management motives is the biggest reason employees, whether unionized or not, fear participating in workplace improvement. Narrow job descriptions are seen as a lifeboat that will see them to retirement—without realizing that the lifeboat has a leak. Management and staff have the same fear as soon as they see that much improvement comes through reduction of their work.

A policy on job security has to be presented in such a way that all employees have confidence in it.

First, the company can try to obtain more work. A great motivator for a work force is to resume production of parts previously sourced out. Cost reductions may shift some make/buy decisions that way, but companies to whom the work has been sourced may also be improvement-minded, so a large volume of work may not boomerang back. Another approach is to aggressively seek contract production. Best is a hot new product, but a banker is not likely to bet on any of these at the time a policy on job security must be affirmed.

Guarantee employment? Not unconditionally. If company presidents cannot guarantee their own jobs, they cannot guarantee the jobs of others.

Limited guarantees are feasible if finances are not desperate. The company will not dismiss an employee outright because of productivity improvement, but only if sales and profits decline; thus people have more than enough time to find a new job.

Formal policies mean less than attitude does. If a flinty treasurer rolls the notes on nervous bankers so employees can buy groceries, they understand that managers (and owners) intend to participate fully with them. In situations of less danger, it is a matter of reading intentions—by all parties. Both managers and employees stop perfecting their attitudes somewhere short of sainthood; but, as in a marriage, a meaningful demonstration of real intentions from time to time makes the difference.

Issues of job security cannot be neatly solved. In multiplant companies, given substantial productivity improvement and flat sales, eventually a plant must close—and perhaps more will follow. How closing is done is what counts—"outplacement" rather than "good riddance."

That follows from attitude: the absence of any great distinction between the policies for staff, management, and "others," and the recognition that operative workers skilled in improvement make the difference between an organization tuned to produce competitively and one that just runs. A company with that kind of work force and an aggressive approach to new product and markets has a better chance of avoiding a graceful plant closing, but first comes the demonstration of company intentions should matters come to a test.

Sharing in the Improvement

One of the first management thoughts is that to be properly motivated, the work force should have some form of gainsharing in the improvements. That depends on what there is to share. In a case of survival, not much. In a highly competitive situation, workers should not be deceived into thinking that productivity improvement is creating a fat pie from which they are entitled to a large piece. Competition forces nearly all the pie to go to the customers.

For a time, better work environment and job security are sufficient reward for workers. Then, if there is something to share, they would like a part of it.

Gainsharing plans come in all varieties, and most companies craft a plan that seems to suit them. Three different types of plans are best known:

1. *Scanlon Plan:* based on estimating the value of productivity improvement through suggestions and group improvements, and covering a total organization.
2. *Rucker Plan®:* based on improvements in value-added, emphasizing involvement, and covering either all or part of an organization.
3. *Improshare®:* based on improvement in standard hours of output as measured by more traditional industrial engineering methods, the value of which is generally split 50/50 with shop employees. Involvement not being emphasized, Improshare is more acceptable where labor-management relations are adversarial.

Gainsharing at a minimum should help break worker resistance to productivity improvement. They no longer will be penalized in the paycheck for it. However, gainsharing is no substitute for making improvement a normal part of a worker's job and for creating the environment in which this can flourish; and a hazard is worker expectations of a sustained high payout.

The earliest American companies deeply into JIT/TQ have not instituted any special plan for gainsharing. Some have profit sharing.

Suggestion systems with payouts also motivate with money. To be effective, they must be well managed, which entails creating an environment where good suggestions are developed and seriously evaluated. Much more than a clever payout plan is necessary, and again the root of problems is the method of management.

Suggestions may point up oversights of supervisors or staff people, often unbeknown to the suggestor. They are often more involved to implement than the suggestor, whose perception of a total operation is limited, can understand. And economic evaluation of a suggestion may appear to the suggestor as a cryptic code. Slow evaluation and a high turndown ratio discourage all but the imaginative few.

Suggestions can be improved by training the work force to improve their workplace, by prompt evaluations and decisions, by recognition, and even by training them to do simple economic evaluations. Over the long haul, the payout for suggestions is best kept low if a large number of people will participate. A flat fee for any suggestion accepted is simple—no fuzzy evaluation of payout delays anything, and no arguments ensue over who contributed how much to the suggestion versus how much to pay out.

Employee participation in ownership is sometimes considered the ultimate motivation, but employees as a whole are not very entrepreneurial and much prefer sharing profits than sharing losses. Ownership is also no guarantee of agreement on management policy. In 1980, "employee-owners" of South Bend Lathe Company, unhappy with wage negotiations, in effect struck against themselves. Underlying this action was unhappiness with their role in management under the ownership plan.

Research on the long-term results of various ways to share results with employees is not very clear. Short-term success can generally be demonstrated, but after the plan has had its effect, then what?

No plan is without difficulty. All exist in the context of management attitudes and methods of implementation. All are surrounded by a particular management's approaches to improve manufacturing. All are subject to circumstances of change and competition in the markets in which a manufacturer competes. No "magic plan" can encompass all this.

Some complexity comes from the assumption that a great gulf should exist between the role of management and the role of the worker, codified by dress and perpetuated by different reward methods and treatment for each. That gulf goes deeply into the culture. Song and script emphasize it.

Think differently. Realistically, but differently. *Accept that workers are to do more than make things: help make them with better quality, in shorter time, at lower cost. They are the technicians of a "production laboratory."* Management is still management, but of a different kind of entity.

In no other culture does anyone seriously believe that worker "cooperation" will match the docility of Japanese. That is not the point, and Japanese workers have not always been "docile" either. The point is to recognize that merely doing a job is not sufficient. Part of *everyone's* job is constantly doing it better, using systematic means.

How to start with workers? By working with them on their work; improving quality procedures, layout, standardization, workplace organization, and preventive maintenance at first; finally, working into such matters as cycle time analysis. Work with them and "treat them right," and eventually people will come around.

Management attitude must accommodate to this. The reward structure follows. Development of skill is a premium requirement. Systems changes enable doing it better. No detail of management system or behavior is untouched.

Some workers will thrive in this; some will not do so well. Same for managers and staff. Workers will not run the company or be permitted irresponsible behavior. Management does have to pay attention to workers to harvest the contributions within their domain to make.

Participation Groups

The most popular name for these is quality circles, but most companies have their own unique name selected for local appeal. These volunteer groups of 5 to 12 employees usually meet an hour or so a week on company time, but sometimes on their own time if they are enthusiastic. They may deal directly with problems concerning quality and productivity, or they might work on something indirect, such as how to control parking space. Each circle has a leader, and the rooms, times, materials, and paraphernalia are generally arranged by a facilitator.

Sometimes these groups work well; sometimes they fall flat, depending on the leadership of the facilitator and the management attention paid to them. Despite all the literature on how to operate them, many mistakes are made. The most common mistakes arise from grossly underestimating the degree of both management and worker preparation necessary. Many JIT/TQ plants in the United States do not have them or have failed with them.

A participation group is no substitute for the daily communication that should exist between workers and supervisors in a JIT/TQ plant. At Hewlett-Packard plants, for instance, supervisors hold a meeting with employees once per shift. These meetings are short recaps of what went wrong, and what went right, and how to do it better tomorrow, unless some special situation occurs.

More apt names for special groups, properly used, might be contribution groups or improvement groups. Time, patience, and training are necessary before these groups can be expected to make sustained contributions. Both the managers and the workers must learn how to work in this way.

Many workers are not interested in participating. Volunteers from 25 percent or so of the total plant work force seems a good average. As in any activity, a minority of the workers will make a great difference if they are well directed, but they take time developing themselves for it. Participation groups are good training forums as well as sources of improvement methods.

Start with experienced workers. They need to understand the company and their jobs to prevent thrashing in unproductive directions. Background in quality and methods improvement is desirable. Even then, groups at first tend to work on trivia that annoy them: noise levels, location of drinking fountains, color of trashcans. These problems are important to them, so pay heed.

The management task is to eventually steer the group in productive directions. A popular thought is that participation groups should work on whatever they choose. That leads to trouble. If they want to change the entire plant layout or revise the quality system, they will run afoul of management and staff who deal with such matters.

At first, the facilitator and group leaders can direct things with questions concentrating attention close to their own work: "Describe your own job, or part of it, using an element listing much as an industrial engineer would. What are the problems you see?" "Describe a problem you touch every day with your own hands." Small-scale problems close to the work are likely to be solvable without being turned into Big Projects, and more quickly resolved, too. A few victories help morale.

A participation group should have management permission to work on a large-scale problem. The agreement is that management will not tackle the problem while the group takes a reasonable amount of time to chew on it. If they propose something, management must review it. If they give up, management goes at it again, and the group moves on to something else.

All but the simplest participation group recommendations generally require some allocation of resources to be put into effect. That constitutes a management review and decision. A common problem is to underestimate the amount and level of management time consumed by this task when many groups are in action. If the groups have not been "brought along" effectively, the time consumed and the reject rate spell a demise for participation groups. Everyone is demoralized.

For that reason, good policy is to not start too much activity at once. Work into it slowly and do it well. Better to have a waiting list of volunteers for involvement groups than to have so much going that nothing can be properly developed.

A considerable start can be made on JIT/TQ without participation groups; and, by contrast, great benefit can come from participation groups without doing much toward JIT/TQ. However, workplace organization and visibility practices are great stimulators of employee involvement in many ways. The progress rate is limited by the rate the improvement effort can be handled and by the need to keep it all in control, coordinated with management and staff work, using participation group contributions. Everyone has to feel their way into this new role. It is not done in a year or so—but rather much longer.

APPENDIX: MEGATECH ULTRA FAST TECHNOLOGY DIVISION (MUFTD)*

The corporate staff delegation from Megatech headquarters scribbled critiquing comments as Division Manager Gary Crowe intoned

*Note: This little larger-than-life apologue is fiction based on observation of many companies. Resemblence of specific names and incidents to any real company or person, living or dead, is strictly coincidental. Read it with an eye contrasting the spirit necessary for manufacturing excellence with the motives herein

his best take-charge sonority, presenting the status of the Megatech Ultra Fast Technology Division (MUFTD). "Third-quarter unit shipments slipped 55 boxes below the second in a soft market, but the mix upscaled to 24 percent VX500, giving us about a $1 million increase in billings and a similar uptick in gross margin, although numbers aren't to the penny yet. VX500 is a real winner, triple our start-up forecast, and the backlog has ballooned to seven months already. A second line is ready to go, but we do not have burn-in capacity yet, and not enough parts. The thing eats 256K bubbles like candy. Purchasing is working with the suppliers like bees in a clover patch, but we can't suck up enough of them."

"VX1000 is still on for a late FY87 launch. Development is hectic, but we should make a plug-in at the Spring Computer Show. As usual, engineering is straining to move on out the curves in scope and speed, while marketing chafes to show what they have already."

"A lot depends on Welco, our primary 256K supplier. We're pushing them before they have stabilized their yield, but Bill Jeffries, their main brain out there, assures me they will stuff it in the sock soon."

"In October, we will close out domestic production of UF32. At $20,000 it is still good for a piece of the action, but Hitachi is dealing for $17,000 to $18,000 now, so we have to make a move because we will not drop below zero margin. Despite our progress in manufacturing, we are not hanging with their learning curve. We can chop almost $2,000 just by doing a full buy in the Far East, and the labor tab in Taiwan knocks off another $400. We'll have to station a couple of test engineers over there a while to respec output with the new components, but we are expecting a fast start-up. The move should keep us within haggling range of Big H anyway, and with the VX coming strong, any layoff should be temporary."

"Incidentally, I bumped into Fujimoto last week in the L.A. airport. Asked him how's business, and he said, 'Not so good. Sales below target, costs above target, and new product behind schedule. Just surviving.' Inscrutable as ever."

---

exhibited, and perhaps it will clarify the primary moral: Manufacturing excellence is attained only in part by technical breakthrough and management technique. The major portion is learning to *live it* throughout an entire company.

"UF32 has been a great training ground for SQC, process simulation, involvement meetings, MicroCost, robotics, and JIT. In 1985, we cut defects by a third, trimmed 11 percent from direct labor, reduced inventory 21 percent, and paid out $41,000 in suggestions. One supplier even delivers daily."

"We think a lot of this can transfer into the VX production, and we are starting some moves there. In addition, the VXs are the first product line fully designed using CAD/CAM and Autotest."

Gary buzzed through the remaining divisional highlights in 10 minutes, concluding with efficiencies, utilizations, market penetration, field service, and a unique three-dimensional cost matrix. Abruptly, he brandished a control wand and flicked the room to full darkness. On the screen, a multispectrum strobe lighting sequence touched off an ingenious 14-minute marketing tape, which combined graphics created by VX1000 with equipment displays and explanations, systematically punctuated by the driving sound of the Ragged Edge, a talented soft-rock group. When he relighted the room, Gary could see that this touch of show business had had the intended mesmeric effect on the gnomes of Rock Candy Mountain, the term universally used by divisionals referring to folk from Megatech corporate headquarters.

The gnomes recovered quickly, tossing accolades on Gary's performance with the division, and Gary instantly sensed that a few feelers for headquarters citizenship might be in his future. Then they took their turn.

Chuck Zorn, Vice President, Electronics Group. Squint eyes marking years pursuing his favorite avocation of flying, blazer and oversize collar betraying an engineer uncomfortable in ties, Zorn praised the status of MUFTD and each of the five divisional staffers present, then power-dived into his message.

"Great still has to become better; and for that, I think we have to promote symbiotic excellence between all divisions in Electronics Group. We have increased the transfers of R&D engineers between divisions. Next we can exchange the best training personnel among us for each of these management programs; and best, this year for the first time we will hold each division's annual marketing meeting simultaneously to catch the spillover effect from the scribbles on the napkins in the evening. We must do a better job matching our technology with the vast applications opportunities out there."

Kyle Morgan, Corporate Director of Purchasing. "We have fewer serious delivery problems with items negotiated through our worldwide sourcing program. Quality problems occasionally, but not delivery. Corporate leverage helps with more than just price. The 256K is one of the items that only MUFTD uses, so it is not in the worldwide purchasing base. But if that item, or any other serious-trouble item were transferred to it, we might help—perhaps not this time, but in heading off future trouble of a similar type."

Linda Stiles, Corporate Director of Quality. "The Megatech Training College is now staffed up for 12 modules in total quality control. I know MUFTD is well established in the SQC ones, but there are seven more, and we can certainly provide a good train-the-trainer package on each for all divisions now, so pick some people to send as soon as you can. Also, we have nearly finished the third one in the series for engineers. It's called, 'Parametric Design by Analysis of Variance.' It can be integrated into CAD/CAM, and it will stop many quality oversights before they are ever begun."

Alberto Corelli, Electronics Group Controller. "I dislike always being the dash of cold water at these parties, but Megatech has a problem. The financial analysts seem to have gotten on a kick of comparing the asset utilization of high-tech companies—who knows why. They pick on our asset ratios, and they like to sift through the 10K reports for evidence of exactly how capital assets in different categories are defined—test equipment, material handling, automation—that sort of thing. So we're doing some work to come up with guidelines for potentially resorting the asset base at each division. You can expect something on this to hit within the quarter because the analysts got on Heilweil's case at their last meeting." (Alexis Heilweil was the personification of the tall-tantrim CEO, who zipped through each Megatech facility at least once per year to keep in touch.)

Clark Gruenther, Corporate Director of Computer-Integrated Manufacturing. "I'm sure all of you have been hearing about integrated manufacturing database management. We haven't been making noise, but something like this cannot be done in a quiet little corner."

With that, Gruenther began unfolding the layers of this newest in corporate management methodologies. It consisted of two major parts, (1) computer-integrated process planning and (2) computer-integrated process control. Everything keyed off computer-aided design, and trickling out from that was computer-generated bills of material, integrated material requirements planning, computer-assisted process sequence, and then there was optional process data collection for labor, material tooling, and quality. The quality system took off from Autotest and would keep quality and yield data sorted in various categories—yields and defect causes in database. The whole thing was tied together by a common data transfer language, and the idea was to whip the problem of systems that cannot talk to each other.

Gruenther rolled through this overview in an hour, pausing only twice to recount favorite war stories from his manufacturing days. Gary Crowe had the MUFTD library update the electronic mail with the snippets of new developments, but he had to admit that some of the concepts loaded in this bomb were new to him, and the fuse was armed for THE DISCUSSION, which followed as the gnomes began working their pre-agenda.

Zorn said nothing as Gruenther and two cohorts zeroed in, beginning by extolling the past glories of MUFTD.

"You were the first with MRP, closing the loop before anyone. You're the only one with automatic forecast updating. I saw my first robot in this plant. CAD/CAM is old hat here; guided vehicles no trick. If any place can do it, you can. We will apply the full power of corporate systems to developing integrated manufacturing database management for this division, and when it has been proven, Megatech will have a package to market that will bring manufacturing right into world-class 21st century status anywhere. You need not say anything now. Just think about it. We'll provide an on-line demo anytime you want."

Meanwhile, Back at the Ranch. One floor below, George Saunders sat just below the Quality-Is-Number-One banner, relaxing his grip on the telephone. "How many will they loan us? Five hundred for a week? . . . Good deal, Charley. I'll pick 'em up myself at Delta Baggage in the morning . . . You'll get them? . . . OK. I'll make sure the records stay straight, and everything . . . Buy you a beer if they're on the line before eight; two if they're here by 7:30 . . . Right! Bye."

George put down the phone, elated as a production manager ever becomes, and called over his evening supervisor for printed circuitboard subassembly.

"Doris, skip the 64K RAMs at insertion on the VX Buff Board, load everything else, and stack them until morning." He and Charley Foss in purchasing had just saved two days' production on a subassembly line, their second such save of the month.

The situation: Welco was late shipping 64K RAM memory chips for two reasons. First MUFTD had stepped up production without that fact working its way through the materials planning system that pumped schedule requirements to the Welco plant. There was a slight planning delay in MUFTD's system, and a bigger one in incorporating the notice of it into Welco's planning system. Second, MUFTD had experienced an increase in boards that had to be scrapped. Experienced operators worked the boards using the 256K bubbles, and both Welco and MUFTD watched that one, each keeping a week or two of buffer stock in hand. However, George had had to put relatively green operators on the lesser skill boards, and the first few weeks, 30 percent of them were ruined beyond repair, so the consumption of 64K RAMs shot up, some of it going into the can.

Unfortunately, Welco had chosen this same time to start up some new equipment on their old 64K RAM process, and their yield had been erratic during the same time. One shipment rushed from Welco had an 18 percent dropout rate at MUFTD on the RAMs themselves, a mountainous spike compared to their recent history on that item.

Welco shipped 64K RAMs according to a weekly shipping release sent by MUFTD, but this week's shipment had not yet arrived due to Welco's internal yield shortfall; and although George had transferred an experienced operator back to the line to stem the excess waste of boards, by Wednesday morning he could see they would not make it through the evening shift without the reservoir of 64K RAMs running dry.

George phoned Charley Foss with a red alert. Charley's prime time was mostly taken with shipping schedules and quality issues involving suppliers. More important, he had just transferred from a sister plant in Orlando a year ago and had connections there, and that plant happened to also use the 64K RAM. Charley first telephoned Welco, who assured him they were cooking a new batch and yield looked great so far, nothing but superior specimens in the mid-week production dedicated to Megatech. On Friday, they would ship a double order to Charley plus meet their regular coverage at Orlando. With that knowledge, Charley persuaded a buddy at Orlando to loan a week's supply of 64K RAMs, shipped by Delta Airline overnight, with repayment the next week by the same means (Charley picking up the freight cost, naturally).

With the 64K RAM loan inbound, George would lose only the ability to completely finish the Wednesday evening shift production; otherwise, he was dead for the rest of the week on the five-person line assembling the VX Buff Board. He had to transfer operators, and he risked shutting down final assembly due to a shortage of a printed circuitboard that was considered light duty in the degree-of-difficulty range in his department—a real black mark. Just as bad, he would finish the week with a wounded earned hours report—a mark that might not be totally black but certainly considered brown and soft, and it was already muckle-dun from new boards, engineering changes, and green operators.

The earned hours report was Megatech's version of the production efficiency report. Calculation of the measure was based on earning a fixed number of labor-hours for each good board sent to assembly, with hours deducted for boards returned from assembly as unusable. That was earned hours, and a simple version of the measure was a ratio of earned hours for the week divided by total direct labor-hours compiled on the time clock (which, in Megatech's case, was really a log-in computer terminal).

This report was the heavyweight item in George Saunders' management report card. Joe Burke, his manufacturing manager, said something about it almost every week. But George never knew whether Joe was half-serious or full-serious with his little remarks: "Hey the hours are rolling in! You must be discharging capacitors in their seats." "Things seem off a tad. Too many breaks for birthday parties this week?" So far as George knew, every pin-striper scrutinized the minutiae in those weekly reports right on up to Rock Candy Mountain and Heilweil himself.

In his 14 years of supervision, George had accumulated comprehensive savvy on every factor affecting earned hours ratios. Other production managers sometimes called him "The Fox." Five years ago, he had deftly planted in a plant manager's brain the idea of adjusting the earned hours upward for time spent on rework, using the justification that such an adjustment made for a more accurate standard. The adjustment factor was based on the best three out of the past six months in percentage of departmental labor spent on rework.

With three successive engineering managers, he had successfully negotiated the following policy on rework. George always phoned engineering if a quality problem could not be resolved within 15 minutes. If the problem appeared not to be human error, and the cause was still unknown at shift's end, it was automatically classified as an engineering problem so the rework hours spent did not count in the earned hours ratio.

A year ago, George had argued for a material shortage adjustment, so that if a supplier did not deliver or if material was declared unusable, the hours lost could be charged to an exception and dropped from the ratio calculations. After a churn through two committees, a compromise was reached. George and other production managers could drop only half the hours lost, on the theory that partial responsibility kept production alert to part quality and to potential shortages instead of just waiting to be served. MicroCost could easily handle mega-category variances.

The shortfall in 64K RAMs was only one of the problems of the week for George Saunders—he had a new component sequencer to deal with. Installed the previous week, the equipment representatives had stayed for two days coaching the operators, and everything seemed to be working. But on Tuesday, the operators experienced component feeding problems, which caused both delays and errors. The representatives returned, and that morning he had spent two hours embroiled in a discussion with them and maintenance on the methods of programming the machine and making adjustments in feed. Meantime, George was employing the old sequencers as backup and trying to keep untangled who was running what.

George's inserters were also overloaded, and on Monday he had brought back two old ones to use as backup on the overload. But this had cramped the space available, and the operators who really knew how to run them were no longer in the department.

The recent increase in schedule had also fouled the flow of material from his department through wave solder, an area used jointly by three departments. Until this week, George had been able to clear VX500 boards through wave solder in the first three hours of the day shift, but now this was interfering with boards coming in from the other departments. When his soldered boards arrived later in the day at inspection, test and inspection people were left waiting for work at midday, and then the evening shift had to stay overtime to finish out.

A test engineer was still working with a new functional test module developed from Autotest. For some reason, it did not perform the same way as several tests already programmed—something to do with changing the programming to increase the speed of functional test. Meanwhile, no one was quite sure whether several boards met the functional test specification—another delay, and another stack of material.

Now that the 3–11 shift had settled in and the crises were reduced to simmering, George had time to review the situation and prepare for the morrow. It was all a pretty normal set of problems, he thought to himself. He had managed to get atop this week's schedule increase by starting a build-up the previous week, stuffing a little padding in the department's stocks, and, as it turned out, he had had to use them. Baring any more unforeseen glitches like the 64K RAMs, he should be through the week with his reputation unsullied and his earned hours intact—a survivor of many similar episodes and one with a potentially higher future at Megatech, despite having finished only a B.S. in an undistinguished evening program at Brevard Tech.

As he was preparing to scroll through the electronic mail and rifle his in-basket, the telephone rang. The voice sounded familiar. "Is Mac Dalrymple in?"

"No," George replied. "You're looking for customer service, but you have George Saunders in production. Do you work here?"

"Not now, but I did. This is Tom Simms. I've been in sales engineering for three years now, and I have a honey of a deal working if only we can make a few changes in that VX500. Twilly Productions could use it."

"Twilly?" George queried. "I didn't know we sold to them. You hardly ever keep track of who the customers are down here."

"We don't, but we could," replied Simms. "All we need is just a little more oomph in capacity of the VX500 to tickle their graphics guys pink."

"Why don't you wait and sell them a VX1000? I understand that is in the works for nine months down the line. Hell, we're just patching together the VX500s now, but I guess it is not mine to decide such things."

"It's a timing problem," said Simms. "Twilly is scouring the universe for something to use for a new production, and if we can't hit them right away, they may not wait, and my foot will be in my mouth instead of in their door. Sorry for the wrong number. I'll dial again. Gotta get on it. Bye."

George resumed sifting the stack from his in-basket, on one side of which was taped a sign saying, "MUFTD is a just-in-time organization," while the other side displayed a yellowed you-want-it-when cartoon. There was a memo initialed by 10 executives, authorizing addition of 15 VX500s to next week's schedule because marketing needed them for demonstrations. George digested that, then made quick work of the advertising brochures and the seminar invitations for everything from project management methods to telephone procedures. A little more slowly, he shuffled through the remaining papers: Notification of an involvement group leaders and facilitators meeting. A plea to find an hourly employee to contribute to the divisional newsletter. A reminder to coordinate preventive maintenance time on all green tag equipment (motors and moving parts, but nonprogrammable).

Next was a memo rejecting Lena Turley's latest suggestion. Lena worked manual insertion on the main VX500 printed circuit-board, and she also sheparded the boards through wave solder. George reread her suggestion:

"Immediately after manual insertion, mount the VX500 boards in the same carrier frame used at wave solder. Stack them on a cart and move them to wave solder and feed the boards directly in. That will save time finding the correct carrier frames at wave solder and it will save double-handling." Signed: Lena Turley.

The response was two pages of estimates and calculations, which boiled down to rejecting the suggestion because it was uneconomical. Sturdier frames would be needed. Why was not clear, but they would cost $8,500—hefty, but shop labor for such things was not cheap, and two new carts would be necessary at $800 per. The time saved at projected volumes would amount to only $14.21 per day to eliminate double-handling, which was a three-year payback, much longer than the two years allowable for such things.

No mention of the time finding the carrier frames and the quality possibilities. The memo thanked Lena for her contribution and urged her not to be discouraged, but to try again. It was signed by one Loren Crowder in manufacturing engineering.

For a moment, George considered telephoning Mr. Crowder to ask him to explain his response to Lena himself, but with a well-practiced profanity, he slapped the memo in his To-Do folder and began to ponder the metaphysics buried in the latest quarterly plant capacity report. A short assembly line had been torn out, so capacity had dropped; but VX500 was still not up to speed on the lines converted for it, so plant utilization had dropped to 66 percent, just above the 65 percent figure considered minimally acceptable by both divisional management and Rock Candy Mountain. Their guidelines were to tear out capacity if overall utilization dropped below 65 percent, and add capacity if utilization broke through 85 percent.

By experience in using Megatech measures, if utilization rose above 85 percent, deliveries began to slip. George had lived through three 85 percent periods at MUFTD. All three times had been travails of burgeoning backlogs, quality quick-fixes, priority-switching, and crash-mode expediting, all accompanied by anguished wails from marketing. Several times George had asked why 85 percent was not shifted in scale to become 100 percent, but no one had ever answered his question.

Finally, at the bottom—to George's disbelief—he found a poster sent by someone in the quality directorate at Rock Candy Mountain, which he was seriously asked to display. It advised one and all that September 8 was the 15th anniversary of the Megatech Zero Defects Program.

CHAPTER SEVEN

Synchronizing the Company

Timing is very important in operations. As established earlier, a repeated pattern of work contributes much to quality and productivity in production and in other operations as well, including those in marketing and engineering that feed into production. In fact, organizing a smooth flow of work there is necessary if a smooth flow is to be obtained in production.

Scheduling and improvement go together. The objective is not to make some magic form of scheduling work, but to *schedule so as to capture maximum possible advantage of regular cycles to improve every aspect of operations*. This begins with the basic approach to the business. A passage from a long-ago paperback, title now forgotten, suggests the opposite condition: "Harbaugh's advertisement read, 'Will do anything legal. Any time. Any place. Fee negotiated.' Not surprisingly he was a little disorganized."

Organizing in this sense is "nonrandomizing" the part of the world one wishes to work with. It is done all the time without much thinking of it, as when developing a set of qualified leads for marketing enticement, separating the susceptible from the random herd. The wasteful effects of random activity are easily underestimated, as is the power of dispelling randomness so that wasteful work is much reduced.

Considerable quantitative modeling deals with the effects of random events. A classic but simple model representing a single truck dock is shown in Table 7–1. It assumes that trucks arrive at an average (or mean) rate, but the actual timing follows a random distribution (Poisson). Load/unload times on the dock average half an hour but are also randomly distributed. (Such models have been

TABLE 7–1 Random Timing of Truck Arrival and Service at a Truck Dock

	Case				
	1	*2*	*3*	*4*	*5*
Time between truck arrivals, mean in hours	2.0	1.0	.67	.55	.5
Truck load/unload time, mean in hours	.5	.5	.5	.5	.5
Percent time dock is idle	75%	50%	25%	10%	0%
Hours average truck waits to load /unload	.17	.50	1.5	4.5	Infinity
Percent time truck at dock only waits	25%	50%	75%	90%	100%

shown to reasonably represent the real thing so long as everyone is just reacting to what happens.)

The percent idle time on the dock surprises no one, but the waiting time of trucks builds up more than is typically imagined. If trucks arrive as fast as they can be unloaded, the line backs up to infinity (were there trucks and space enough) because any small delay in unloading creates a backlog that can never be caught up.

At the extreme, this model turns into silliness. A management confronted with such a mess must act. It can build more docks, or it can increase the load/unload rate on the dock, but the most interesting action it can take is to dispel the randomness in the situation. Suppose trucks arrived at a constant half-hour interval, and each one was unloaded in exactly half an hour, as if the dock operated as a station on a conveyor line.

This is the idea behind synchronizing activities, not just in production and transportation but anywhere possible in the company. Far too much energy goes into work that just reacts to randomness, and too little into filtering it out, both in management practice and in academic modeling of management problems. Synchronizing operations is much more than just aligning a pattern of numbers in a schedule. It attempts to remove as much variance as possible from the process of serving the customer, beginning with the planning of sales and ending with delivery to the customer. Many management innovations come from thinking this way.

Benihana of Tokyo organizes restaurant patrons into a regular pattern for service. Reservations attempt to space customer arrivals throughout a rush period, and walk-ins are worked in if they can be accommodated. Some reservations arrive late, and some wait for others to join their group, so despite the attempt to level-load, arrivals must be held and marshalled in groups of eight in a small waiting lounge. Irritation by the delay is usually assuaged by one or more high-markup drinks.

Each group of eight is led to a hibachi table, where they are treated to a show-biz performance by a chef—elapsed time, about 40 minutes. By the 50-minute mark, a barren table suggests the show is over, and the group leaves, most never detecting that they have been very politely marched through a one-hour episode of a repeating show so captivating that it is much imitated. Their method makes maximum use of chefs, tables, space, and time—not to mention bartenders.

Benihana of Tokyo does more than just organize the kitchen, isolating it from customers. It organized the business for its synchronized approach to service, refining all the details until they fit its "strategy." Something like that is necessary for synchronizing operations in a manufacturing company.

SYNCHRONIZATION AND UNIFORM LOAD SCHEDULES

Synchronization is matching repeating cycles of activity. Cycle lengths may differ; but, if all are multiples of one another, they start and stop together, forming a harmonic pattern of work. The definition is as simple and as abstract as the definition of time itself: hourly, daily, and weekly cycles all fitting within each other.

Production cycles may be as short as fractions of seconds per unit (as in chunking out beer cans) or as long as weeks and months (building jumbo jets). Business activity cycles may also be long or short—days, weeks . . . years—but are tied to some repeating cycle. Business time cycles are so fundamental that, once adopted for a given organization, they tend not to be further questioned.

Regular timing dispels "randomness" from the void of the universe. It does so in a physical sense, and it does so in a business sense. If any human activity is executed with excellence, by no accident is it commonly described as "running like clockwork."

Balancing Production Flows and the Cycle Time Concept

Do as much work as possible so that its time cycles mesh with those of the total activity of which it is a part. At its most abstract, that is the cycle time concept introduced in Chapter 4, but there are several aspects to it:

1. Duration of work is one type of work cycle—time between setups or time between tool changes. Duration should be limited by the number of work cycles until a process capability begins to fade from control. Setups, tooling, lot sizing, and quality checking all may be based on a natural quality cycle. Several processes linked together may key on the quality cycle of the dominant process (Chapter 3). On the other hand, any operation may go through numerous setups having the purpose only of transmitting a uniform load of material requirements to the next lower operations—as is true of assembly operations.
2. Work studied by the cycle time concept should see a decrease in the variance of cycle times. With reduced variance in both work cycle times and duration (setup) cycles, operations can be more closely balanced in operating rate. Fitting the support activity into a pattern of work starts to become feasible (Chapter 4).
3. There are two major cases in balancing running rates:
   a. *Variable work cycle operations* (usually manual): The desire is to increase production rate by adding people and to decrease it by subtracting people. Doing this has two requirements:
      1. Flexible, multifunctional workers
      2. Layouts for flow and flexibility. Allow workers to easily move from station to station and permit them to adjust elements of work content from station to station. This is very important in assembly areas and is desired in fabrication as well. (The simple cell lends itself to this, and it is an outstanding feature of a well-developed U-line. Refer to Figure 5–1.)
   b. *Fixed work cycle operations* (usually machines): Induction heat treating is an example of a process work cycle that cannot vary and still produce quality parts. For this kind of operation, have the machine idle between work cycles, as illustrated in Table 5–1 and Figure 5–2.

Attaining major improvements in quality, layout, work force flexibility, visibility, and all the rest make uniform load scheduling easy. Without improving the process itself, no scheduling method can duplicate the overall effect.

Balancing cycle times is simple organizationally. Each suborganization studies what it needs to do to mesh its work cycles with the cycle times of part use that the overall schedule demands. Whatever it can do, it does, depending on the degree of its attainments so far. However, people must be well oriented to quality, timing, and improvement to do it, and constantly thinking this way is a mindwrench at first.

If production can be organized repetitively, cycle time thinking can be greatly refined; but if the product line does not lend itself to repetitive processes, what then? Consider the job shop one big "variable work cycle operation." That is, if the work force and layout are as flexible as the nature of the work will permit, scheduling and planning are easier. Establish as many visibility methods of shop control as possible, and make use of repeating time cycles in any way possible, however limited.

The Uniform Schedule

In the repetitive case, an overall pattern of work is set by a final assembly schedule. Having the use rate of all materials throughout a specific production period is the easiest possible pattern of work for each supplying operation to match—perhaps not perfectly, but as close as possible. Thus the short definition of a uniform or level schedule is, "Make a little bit of everything every day."

Setting up a production schedule for uniform usage over a specific period of time entails more than a technique for final assembly scheduling. Activities that feed production must mesh with it:

Marketing
- Sales planning or demand management.
- Order entry and customer promising.
- Logistics.
- Field Service.

Engineering
- Design of product.
- Timing of changes.

Production process

Design and development of process.

Cycle time concept.

Visibility. . .

Processes of *suppliers*.

Scheduling

Based as much as possible on simple linkage of synchronized activity, and as little as possible on managing inventory that separates uncoordinated activity.

For the assembled product, material usage at final assembly draws material into it if a "pull system" of control is used, so scheduling is a preplanning of volumes and sequences in final assembly during a production period. Scheduling must also advise the supplying operations of the approximate volumes and sequencing so that they can plan to mesh with final assembly. The actual performance may deviate a little from this game plan. The intention is to prepare for as much synchronization as can be mustered.

An Example: 3M Magnetic Media Division, Hutchinson, Minnesota. The 3M Magnetic Media plant makes magnetic recording tape for home video recorders and for commercial use. Much of the production consists of half-inch VHS video tape made to order in a tough, competitive market. There are about 100 different end-item variations of this product, not counting the different labels applied for different customers. Only 30 to 40 variations are commonly called for. Variations are in the configuration: different cases, cartridges, tape lengths, speeds, and so on. All are made from the same material from a coating process, which is the technical heart of the plant. However, delivery problems came from the downstream operations that slit the tape and converted it to a final form for the customer.

In 1982, 3M Magnetic Media, Hutchinson shipped only 40 to 50 percent on time in any week. Poor delivery of a cost-competitive "commodity" product was unacceptable if 3M wished to stay in the market. For about two years, it went through the soul-searching process, finally deciding that it had to go to JIT/TQ to survive.

In late 1983, it began holding meetings of managers. The meetings gradually began to include more and more people. One of the critical items was how to rethink the scheduling process. That, they soon discovered, required rethinking the production process and the marketing order entry process. Product engineering changes were not much a factor in this case.

A production period two weeks in length was decided upon, somewhat arbitrarily. Management believed it could get through 20 to 30 different varieties during two weeks if setup times were reduced; and, if so, it should be able to produce a mix close to that demanded by customers for two weeks. Marketing was apprehensive but agreed to "freeze" schedule for two weeks. Very few genuine emergency requests from customers could not wait two weeks. Many expedite requests were caused by customer panic through late shipment.

Not every variation of half-inch VHS tape was produced every two weeks. High-volume repeaters were in every schedule along with a mix of "occasional oddballs."

Substantial revision of every detail of the production process was necessary to run by such a schedule. Production would operate by the packing process, "pulling" material in from the feeding operations—a simple form of "Kanban square" system, which they had to work out for themselves using their own order sizes, cart sizes, and cassette sizes.

The details could not be masterminded. They came out during Saturday simulations using styrofoam cups on brown paper layout models of the operation. All supervisors and key production personnel were involved, though not all at one time.

As a result, they decided to run by a daily schedule those items of sufficient volume that one pallet-load or more could be packed per day. Lower-volume items were run as complete orders. Finished packing was shipped to a customer as soon as enough was accumulated for economical shipping—or a customer order was complete.

Layouts changed. Quality delays were addressed. Equipment and tooling modified. Locations and containers specified for the workings of a "pull system."

About five months after the first Saturday simulation, the new scheduling system turned on. The agreement with marketing was that anything in the two-week schedule was as good as shipped, and do not promise a customer delivery that forces breaking into schedule unless they have a *real* problem. Priorities in the subsequent two-week schedule are at the customer's service.

In the first two-week period, 3M Magnetic Media, Hutchinson shipped 95 percent correct items to schedule. (Both overshipment and undershipment are deviations from schedule.) For the following year, it was 98.5 percent to schedule, and the major cause for late shipment was unavailability of the customer's label. (The company is working on that.)

Marketing's fears subsided slowly. They prevailed to break into the schedules only about once a month to provide something special. Service improved, but marketing wanted better.

Work-in-process inventory decreased, and finished goods remained low. (Previously, shipping quantities were cumulated also.) The personnel and the work previously required for detailed scheduling and control of the conversion shop floor disappeared. The internal materials planning system simplified, though it must still provide for long lead time materials from suppliers.

After a year or so, the schedule period was shortened from two weeks to one. Then 3M Magnetic Media, Hutchinson could within a week regularly deliver small orders and begin delivery of large ones, but this response time was still considered too long. Marketing wanted same-day service for at least limited quantities of any variation of half-inch VHS tape.

Rather than stock all the variations of tape cassettes, production believes it has become flexible enough to provide the equivalent of off-the-shelf service. Now extra time is provided in each daily schedule to run small orders for same-day shipment. With experience, small differences in cycle times at each operation for each daily mix can be overcome through worker and equipment flexibility.

Development of a Uniform Load Assembly Schedule

The development of a uniform load schedule for a situation somewhat like the 3M Magnetic Media, Hutchinson plant is shown in Table 7–2. The numbers are simplified for easier reading, and the schedule period is a month rather than two weeks. For more complex products running at slower rates, the uniform consumption of parts demands mixed model assembly; that is, instead of completing production in container sizes of 100, as in Table 7–2, it is completed in lots of 1.

An example familiar to many is automotive assembly. Vehicles of the same exact type do not follow each other down the line.

TABLE 7–2 Breakout of Repetitive Items for Uniform Schedule: One-Month Schedule Period

Item	*Volume for Month*	*Made per Day*	*Made per Week*	*Made Per Month*	*Average Daily Volume*
A	29,000	1,450			
B	12,000	600			
C	5,000	250			
D	2,000	100			
E	1,000		250		
F	800		200		$\frac{50,850}{20 \text{ days}}$ = 2,543 on average day
G	400		100		
H	300			300	
I	200			200	
J	100			100	
K	50*			50*	
Totals	50,850	2,300	550	650	

Daily Schedule for Week 1

Item	*Day 1*	*Day 2*	*Day 3*	*Day 4*	*Day 5*
A	1,400	1,500	1,400	1,500	1,400
B	600	600	600	600	600
C	200	300	200	300	200
D	100	100	100	100	100
E	300				
F			200		
G				100	
H					300
I					
J					
K					
Totals	2,600	2,500	2,500	2,600	2,600

*Container size is 100, so the lot size for assembly is 100. This is a half-container exception.

Each unit may have some degree of uniqueness, but the total mix on one line is considerable: two-door, four-door, sedan, station wagon—even different nameplates. That mix is common. Less common, but still practiced, are mixes of two-wheel and four-wheel drive, cars and pickup trucks, and different frame sizes.

In addition to the mix of basic models, automotive assembly complicates the sequencing problems with mixes of different options: engine types, air conditioning, trim packages, and the like. Handling this taxes flexibility of line layout, assembly equipment, parts sequencing to assembly, and the flexibility of the work force. If the assembly process is designed so work can easily shift between stations, the problem is much simplified. (The objective is not a complex line-balancing algorithm, but to remove as many balancing constraints as possible with little or no cost. Work content balancing through mobility is not presently helped much by robots because they cannot move from station to station as easily as people do.)

Similar situations exist with large appliances and large mainframe computers (where part of the problem is production of software options). Two of the toughest situations are aircraft and large trucks. Given that, a simple explanation of mixed model assembly presents only the essence of the problem. Instead of bunching models of highly similar type for assembly, build them mixed-model (as shown in the example below) for a proportional mix of four As, three Bs, two Cs, and 1 D, repeated throughout a production period:

A–B–C–A–B–A–B–C–A–D

This sequence can be arranged several different ways and still come out with the desired result, spreading the consumption of parts for each model evenly throughout the production period. A mix as simple as this is a back-of-an-envelope calculation, even when volumes for each model do not come out in simple proportions. If operations are being developed for balancing by the cycle time concept, scheduling becomes easier. If not, no scheduling magic will make waste-reducing physical changes.

However, at best, options and suboptions associated with each model add more levels of material-spacing complexity. In practice, maintaining a uniform load of the major model components is usually inimical to good spacing of low-use options. Operations producing high-percentage-of-use components see a uniform loading, but those producing the rarely used options see a schedule of "random droppings," the opposite of the effect intended.

If products have not been designed for production, attaining this kind of schedule (and production process) is hopelessly complicated. If options have been designed for interchangeability and layered build-up, an elaborate mix of product for the customer can be handled this way. It depends on whether the company has created "nonrandom order" out of the chaos in its marketplace.

For engineering to design products for a highly meshed production process, it must have that type of process in mind all along. In addition, willy-nilly scattering of engineering changes into a uniform schedule rapidly makes it nonuniform, so engineering changes should be grouped to avoid being random ripples cascading through a process. In a plant like 3M Magnetic Media, Hutchinson, this is not a matter of much concern. In complex, highly optioned production, it is.

The production process is one major reason nonuniform patterns of parts consumption exist even if a schedule for it has been created. If assembly does not proceed at a steady rate, the pattern is lost. If units must be taken out of sequence in assembly for quality reasons, the pattern is lost. If a supplying process cannot deliver quality material on time, the pattern is lost. A revamp of the schedule disrupts the mix of material demanded for every other process supplying the same assembly.

Synchronization has different degrees, and therefore the degree of precision required in developing and holding a uniform or mixed model schedule differs. If processes feeding assembly do so only in daily quantities, then the mix sequence during the day means nothing in further providing a pattern of work to the feeding processes unless they are being developed for closer than daily synchronization. If an auto company has special seats for each automobile fabricated and sent to assembly in vehicle sequence, then seat fabrication and auto assembly need to be synchronized to the unit.

Why do such a thing? In that instance, seats are bulky and subject to damage. Each time they must be handled and sorted for use, the possibility of damage and error occurrs. Besides, all the extra sorting and handling is waste.

Reasons explaining those cases in which synchronized production is difficult or impossible, even in elementary form, fall into three categories:

1. *Engineering* of the product is unsuitable. At the extreme, a totally custom-designed, unique product is only put together once.
2. *The production process* is not developed for it, either not designed for it at all or "quality" problems of various kinds plague it.
3. *Marketing* demand is irregular, maybe because the market itself is very unpredictable—possibly low volume—or because the sales planning, order entry, and delivery are not developed to promote it.

The market may really be unmanageable for some companies; but, for many others, management does not recognize its potential for creating order from random customer events.

Linearity of Schedule

A uniform load schedule is of little value unless attention is constantly focused on executing it. *Linearity* is execution at a steady rate of the prime schedules (usually final assembly) that everything else feeds directly or indirectly. Development of a uniform load schedule establishes planned cycle times, and executing linear to schedule maintains the cycle times with minimum deviation. The JIT company soon learns to establish constant, visible feedback to workers on the "real time" status of final assembly completions to the planned schedule.

In synchronized fabrication and assembly, a common practice is to work only two shifts, using the time in between to catch up should things fall behind. The interval between shifts is also useful for preventive maintenance, trials of process improvement ideas, and trials for new product tooling. The time can also be used for improving the skills of the work force—formally and informally.

Not all companies have fabrication and assembly. Some have processes that must run three shifts. But any company striving for continuous improvement should strive to plan and execute a schedule which (1) establishes and holds cycle times necessary for synchronization and useful for studying improvement, and (2) keeps a little capacity in reserve for catch-up, self-renewal, and improvement.

In fabrication and assembly, the fear sometimes is that not having a third shift reduces effective capacity. Short-run it does. Long-run, two shifts working on constant improvement will be well ahead of three spending no time on self-renewal.

MATCHING THE UNIFORM SCHEDULE TO THE MARKET (AND VICE VERSA)

Where to start? Physical changes and people development are necessary, but true JIT *flow* production cannot occur without the schedule for it. Developing this is less a scheduling method than learning how to operate the company in a different way. Organize the random patterns of customer demand into periods of uniform schedule, each different from the last one. Perform this organizing process in as short a lead time as possible.

1. *Production.* What are necessary entities to schedule to provide a basis for capturing the potential for repetitive activity—for setting the dominant "drumbeat" from which many operations can blend in with their own rhythmic variations—the cycle time concept.
2. *Marketing.* How can we remove unnecessary influences that cause erratic transmission of market demand into the production process?
3. *Matching.* How can the flow of market demand be matched with useful periods of uniform load production schedule?

The schedules to be uniform loaded are typically final assembly schedules or something approximating them, and conditions of schedule development are in three categories:(1) make-to-stock, (2) make-to-order, and (3) make-for-another-manufacturer (often a variation on make-to-order).

Many companies combine all three cases in developing the uniform load schedule for a single operation, so the situation can become quite complex. The objective is simplifying this confluence of factors in any way possible.

Make-to-Stock

The planning for distribution inventory should be on the same planning cycle as the length of the production periods, so production planning always takes place using the freshest possible estimates of need. That is, if the production period is two weeks, replan the demand for finished goods in two-week cycles, completing the fresh "inventory" plan just prior to devising the uniform production schedule.

Plan the distribution inventory to cover specific periods: Lot size for period coverage demand of all items to be produced. (Not every item in stock may be produced in each production period.)

Avoid coupling reorder point systems to uniform load production scheduling. These create irregular lot sizes sent to production planning at irregular times, and they will make this type scheduling impossible or nearly so. The so-called savings from economic lot sizes is illusory. Most such lot sizing is done by assuming that items in the product line are independent in their demand and that any item can be reordered at any time. This aggregates the normal random pattern of demand over different "random" lengths of time and passes it to production. Think *time coverage* over equal lengths of time to marshall demand and match demand patterns to schedule periods. Use the finished goods inventory to perform its function of buffering random swings in demand from the schedule.

A very useful form of finished goods management is called "sales replacement" or "demand replacement." Consider the finished goods as stock to meet demand until the schedule can be reworked the next time, and base the stocking level on that. The amount to be produced is whatever was sold last period plus any inventory adjustment considered necessary for safety purposes.

Items having mild upward or downward trends are handled by "sales replacement" with little ado, and the objective is not to create artificial blips in demand unaware. Have field stock locations reorder stable demand items by "sales replacement" also. That is, they should order to cover recent sales without increasing inventory to provide for a possible permanent uptrend in that item. Field ordering policies should transmit through the distribution channel pretty much what has been sold. Try to prevent overreacting to random swings in demand.

A forecast is necessary for all items not having stable demand patterns, so try not to make an unstable demand pattern where none exists. Seasonal items require building to forecast. So do those subject to demand fluctuations from the company's own sales promotions and from actions of competitors. Some methods of mechanical forecasting are more accurate than others for a given purpose, but none work as well as trying to identify and take into account known causes of demand change.

For this, a good relationship with marketing is necessary. If marketing folk are inventory minded, they will tend to forecast high so as to have a full wagon to sell from. Converting them to thinking time coverage and production response is no small task but a necessary one if they are to provide estimates of demand change that are true deviations from random variance in a demand pattern. In a company stocking thousands of items, this might be a small percentage of the total items.

The idea is to operate the uniform load schedule from as close-in and stable a position as possible. Then, attempt to run as much production from the uniform load schedule as possible.

Planning ahead for capacity and for long lead time suppliers takes more forecasting. That difference is important to keep in mind. Trends up and down and market-caused changes must be anticipated further ahead than just the upcoming production period. However, if the planning horizon for materials commitment can be shortened, reliance of company operations on long-term forecasts can be reduced.

After they get rolling, make-to-stock companies can reduce finished goods stock levels, but an *increase* in finished goods is not unusual in the start-up stages. First, the production operations may have some interruptions to make adjustments and improvements. Second, a company may justifiably not have all the confidence in the world at the outset. Only after establishing production's ability to operate responsively to uniform loading will this confidence develop to a point where finished goods levels will decrease. The reason for the reduction is that production enters finished goods more as a steady stream rather than in large lot sizes. Much of the finished goods remaining is safety stock and pipeline stock.

The ultimate objective is to provide the equivalent of off-the-shelf delivery from production—no finished goods inventory. Flexible, short lead-time production may actually provide better service than if inventory is stocked. If production has a long lead time, any surprise stockouts cannot be quickly replenished, and customer service suffers anyway.

3M Magnetic Media, Hutchinson is an example of a production process well down the road toward this effect. Because of the daily uniform loading, production every day is working in the service of many different customers. Customer orders no longer wait so long in a backlog for their turn to be serviced one-at-a-time. Flexibility is much improved.

Make-to-Order

Synchronized make-to-order production seldom incorporates custom design. That is job shop work—too little potential to capture repetitive patterns. Synchronizable work is assemble-to-order, often with option mixes to contend with.

One example is automobiles. A symbiotic relationship exists between auto manufacturers and their dealers; but yes, from the viewpoint of schedule construction, autos are built to order everywhere in the world. Some orders are one-of-a-kind combinations; others are fleet orders for hundreds of identical units.

Another example is Singer, who builds consumer items differentiated by trim and logo for the Sears and K Marts of the world. Customers want delivery in either large or small quantities, sometimes in response to demand from sales promotions. A third example is Briggs & Stratton. Their small engines have a seasonal demand pattern and are modified to fit hundreds of different applications by the equipment manufacturers to whom they sell.

Uniform load schedules in the make-to-order case are developed from the backlog of orders, but where do these orders come from? Sometimes they come from orders to replenish inventory—the customer's inventory. If the customers understood how *they* might benefit from doing so, their ordering of numerous items might be done on a "sales replacement" basis in time cycles compatible with the manufacturer's production periods.

This example does not sufficiently cover the make-to-order situation. Automobiles are an example of orders being a mix of (1) "sales replacement" orders for dealer stocks, (2) dealer orders based on their anticipation of local demand preference changes, and (3) vehicle orders demanded by users themselves—both individual and fleet orders. The waiting time of all customers must be competitive.

A very uniform schedule could be constructed from a large backlog if customers expecting delivery when promised was not a problem. Customer orders should be assigned to time segments according to delivery promise and made during a production period matching that time segment. Even so, a large backlog is a long wait for the average customer, so the time segments should be short as possible, and the preparation time for them as short as possible.

Preparation time is necessary before a segment of uniform load schedule can be executed. First, orders must be entered into segments of time, or "buckets," which correspond to the production periods. They must be confirmed. Can the customer pay? Is the order correct? Once a bucket of orders is as firm as marketing can make them, they are transferred to production.

Before production ever receives the new bucket of orders, marketing should have selected orders and quoted deliveries in such a way that production does not have an impossible task aligning them into uniform load patterns. Then final planning begins. Personnel transfers, tooling preparations, or even layout changes must be planned. Preparation to run a schedule segment takes place during the periods leading up to it, and finally the schedule is executed at the correct time.

A rough form of this procedure is sketched in Figure 7–1. First, marketing and production have to learn how to basically operate this way. Then, if marketing can become sharper marshalling demand and if production can become sharper preparing to execute, lead time to the customer might be reduced. Schedule periods might become shorter, one less period might be needed to prepare for production, or orders might transfer from marketing daily instead of weekly.

This process can be complicated in many ways. Suppose a company has five plants, each capable of producing the same line of product except for a few details of optional configurations. The transport distance and the option content restrictions must be managed in parceling out the orders in time segments to each plant, and, in addition, orders will be traded between plants unless forbidden. Keeping it simple requires concentration on eliminating the possibilities that make it complicated.

One myth of this process is that established schedule segments are "frozen" inviolate; that is, once marketing has handed the backlog to production, no order changes are permissible. A heavy degree of discipline is necessary, but some changes are always permissible. If a customer cancels, the order must be dropped. In desperation cases, an order can be added; likewise, changes to orders

FIGURE 7–1 Development of Uniform Load Schedule: Make-to-Order Case

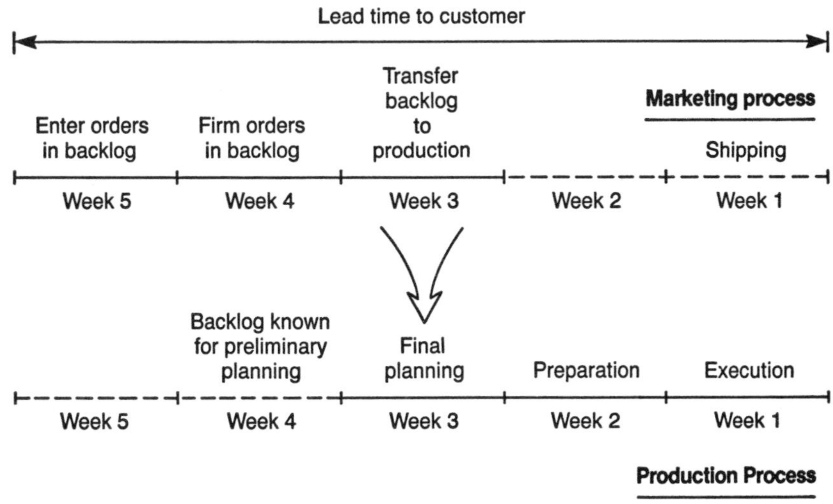

Lead time to customer

Enter orders in backlog | Firm orders in backlog | Transfer backlog to production | | Shipping

Week 5 | Week 4 | Week 3 | Week 2 | Week 1

Marketing process

Backlog known for preliminary planning | Final planning | Preparation | Execution

Week 5 | Week 4 | Week 3 | Week 2 | Week 1

Production Process

can be made—but not very many. Marketing must have enough discipline to protect the integrity of the production system by refusing to disserve the many customers on behalf of a favored few.

How long a schedule segment? Long enough to allow synchronization of operations through which material flows, but how long that must be depends on the status of the production process for quality, dependability, and flexibility (Chapters 3 through 6). Over time, an objective is to develop a production process that can quickly respond to *real* changes in market rate and mix, but the development takes discipline.

Short lead times make this discipline easier to achieve. A large backlog and long lead times invite customers and marketing to "play games" for a long time before they become serious. Large backlogs are loaded with orders that will never be built: quotes never accepted, possible orders never completely defined, "maybe orders" from customers just waiting in line because they might want something.

Managing a close-in backlog is necessary for marketing to help align orders to time segments in a preliminary way. Several other kinds of considerations can affect the timing of actual production. Two of them are:

1. Credit terms causing orders to bunch instead of spread. If customers have, for instance, until the 10th of the month following delivery to pay, by no accident will many of them want delivery in the first week of the month to maximize time before payment.
2. Transportation planning often "lumps" production of a given type. Items must all meet a particular train or boat. In some cases, shipping times must be coordinated with other plants (peripherals for a mainframe computer are one example).

Discounts, promotions, selling practices, reporting periods, credit terms—all may reward customers for large lots shipped all at one time or for long-term orders made and shipped as needed. With this concept in mind, the search for transport economics shifts from hauling large lots at irregular times to hauling mixed loads at regular times. Marketing then beckons customers to work within a smoothly flowing system rather than disrupt it.

Developing a uniform load schedule specific to the business situation is one of the most difficult aspects of JIT/TQ manufacturing. The situation at the 3M Magnetic Media, Hutchinson plant was simple enough to briefly describe but complex enough to tax the people involved. In the beginning, they could not lay out the steps in a PERT chart but had to piece it together much like solving a jigsaw puzzle.

Make-for-Another Manufacturer

Supplier companies usually sell to more than one industrial customer, each presenting demands using their own systems. Should these customers happen to operate by uniform load schedules themselves, scheduling by the supplier becomes a matter of matching the production periods and preparation lead times of the customers with the production periods and planning cycles of the supplier.

That issue itself is underestimated. Blending together a set of out-of-synch cycles is a challenge. The discrepancies in time phasing can only be covered by inventory in some way, and they complicate planning.

Many industrial customers will only send orders. Some will also send a forecast of orders to come. Organizing this backlog into a uniform load schedule does not differ in principle from organizing the backlog of orders from nonmanufacturing customers. Much depends on how stable the short-term backlog is, and whether it permits blending orders into a uniform load schedule. In turn, that depends on the quality of customers' operations. Suppliers generally have minimal influence over that except in the way they select customers.

The reasons uniform load scheduling is so difficult include not only confusion over sources of disruption to uniformity and how to remove them but also the nature of customers. A company selling to any major customer with a "big inventory" mentality is unlikely to suggest that the customer change its buying practices as one of the first moves in JIT/TQ.

The challenge is shaping the environment for simplicity. Only by concentrating on that in a unified way can the company (production, product engineering, and marketing) mold the complexity into a simple pattern. Making the complicated look easy is mastery of it.

CUT LEAD TIMES AND SIMPLIFY SYSTEMS

Large inventories and long lead times are both associated with complex systems, and complex systems are both harder to understand and harder to change than simple ones. The workings and ramifications of simple systems are easy to understand. Omark Industries' Oroville, California, facility quickly converted a simple reorder point system for finished goods to a time-coverage system to better uniform load its schedules. Both systems were easy-to-program procedures used in the same room as the inventory.

Complex real processes must be represented by complex systems, but not all complex real processes need to remain that way. Simplify reality and the system representing it can also simplify. Changing a computer system makes no change in reality—transport, tooling, location, layout, work method, or even use of the product by customer—unless tangible activity also changes. This principle is important for continuous improvement thinking, and believing otherwise is a common sinkhole.

Administrative people working with a system far from the activity it represents see only symbols of reality, not the reality itself. Should they begin to confuse the two, they will believe that if a system to "optimize" shop operations is devised, surely real improvement will follow. Improvement changes suggested in this way may go in a helpful direction, but they do not execute themselves, and the programs require data for the remote control.

Real improvement consists of straightening and simplifying the value-adding operations themselves, pruning unnecessary activities in the process. Why not do as much as possible by viewing the reality directly—using the activities themselves as their own system? (Visibility systems do that—or come close.) All staff should deal as directly as possible with real operations, real product, and real customers. They should be aware of the visibility principle and use it, even when developing computerized systems. Then they can snip the tentacles of old, complex systems and devise newer, simpler ones.

Control of operations is also confused for continuing improvement—control by recording where everything is and what is done. Auditable transactions require discipline, and in-control is preferable to out-of-control larceny and confusion. Beyond this, control is not improvement. *Disciplined management of waste is still waste.* Improvement by remote control suggests that a number is out of line and, if no more is known of reality, recommends that "Somebody do something."

Simplification of systems comes through abandoning activities once thought necessary and eliminating the lead time associated with them. Operations, inventories, inspections, and delays that no longer exist no longer need to be planned and controlled—neither scheduled nor budgeted nor reported. However, in transition, the existing complex systems still expect data feeds from these severed appendages to keep going, and without some system capability, essential planning and control of residual activities are endangered if new systems cannot function yet.

This possibility is harrowing even for those whose view of reality is not obfuscated by system. Every company must have some systematic means to define the needs of customers, plan future operations, and assure that assets are as stated. The objective is to do this without waste, but escaping from complex systems is, for many, not a simple matter. Relapsing into a complex system to manage waste is an easy way out.

PRODUCTION PLANNING AND CONTROL SYSTEMS

Many companies call these *materials systems* or even *inventory systems*. Their primary planning and control of production is through materials planning and inventory status, but much more is necessary for smoothly-flowing production: tooling, maintenance, material handling, and personnel, among others.

Rolling that all together into one big, interconnected computer system for a plant of any size creates a monster system. Such a system is technically possible with good preparation, but expensive; and such systems are charming only if the plant has few problems of quality, maintenance, shortages, and the like.

In practice, materials requirements planning (MRP) systems are tailored to fit the circumstances; but in the professional jargon, all of them explode demand for parts and materials through successively lower levels of bills of material, using time-offset scheduling, and adjusting each part quantity by netting out the number still held in inventory. Significant errors in bills of material, inventory balances, or order quantities create a garbage-in, garbage-out situation, so these systems require accountancy-level attention to clerical accuracy.

Most companies working their way into JIT/TQ are not in a position to construct a monster system. The majority have a history of some version of MRP. Changes to the system depend on the status of MRP development when the company began a new direction and on the degree of evolution since then.

MRP systems were originally established to model the company operations somewhat as they existed at the time of MRP installation. If the basic overview of materials flow changed little, the basics of the MRP system changed little. Inductoheat needed no major MRP programming surgery. The company remained a job shop, the bills of material were simple from the beginning, and the computerized version of the system needed no enhancement. The 3M Magnetic Media, Hutchinson plant developed a simple MRP system *after* it developed a uniform load schedule. The key is good discipline with a simple system.

Job Shops

A true job shop must plan and control irregular-quantity, irregular-timed jobs using job orders without any pull system. Simplicity can be attained in several ways:

1. *Minimum levels in the bills of material.* Fewer stock levels to plan and control, fewer part numbers to manage, less time for system processing and maintenance, and fewer systems transactions to manage. To accomplish this, the job shop must attain visible control over its work-in-process inventory to substitute for moving material in and out of inventory. Visibility helps control even long routings through many operations.
2. *Simple lot sizing.* Making only what is needed (lot-for-lot) is a good formula. Lot sizes in nice, round, easily counted numbers is another. If material must be built ahead for stock, lot-size for time coverage so that, for instance, once a quarter every member of a given family of parts is built. That procedure begins to capture whatever repetitive potential is present.
3. *Make maximum possible use of direct visibility.* This minimizes the need for elaborate shop floor tracking by remote system. It also permits easy tracking of a schedule by those responsible for the "support" activities: tooling, maintenance, material handling, etc.
4. *Do not overload a shop with work that cannot be done.* If a shop has a flexible work force and flexible equipment, capacity planning is simplified. Do not overload total labor-hours or total machine-hours and the work will probably get out. Keep work-in-process small and expect prompt attention. Large WIP complicates systems and destroys workplace organization. Make priority choices on work *before* it goes into the shop, not while it is there.

Inductoheat is a small job shop in which control by visibility of the main assembly schedule is very effective. Larger facilities must think more deeply about it, but visibility methods can be devised once minds are set toward them. Space for queues waiting at machines can be marked and limited. (No space for it, do not move it there.)

A very effective shop floor visibility system for material control was observed at a huge plant of Hitachi Heavy Industries. The jobs were large-scale metalworking: big turbines, reactor vessels, and so on, and all were custom designed. Each work center kept work orders and shop paper for jobs in queue in file drawers of a cabinet with easily removable drawers. The job in the top drawer was the next one to work on. When it was finished, all drawers moved up a slot. Paper for a job to be moved into a work center was placed in the lowest empty drawer. If the lowest drawer was already full, no material was moved in.

Each work center had only as many drawers as could hold a few days' work—effective queue control. All workers knew that a job in the bottom drawer should be worked out by week's end, for instance; and a work center starting to bottleneck rapidly became apparent to anyone feeding it, and their duty was to stop and help offload the overworked point if possible.

Flow Shops

A company characteristic of a flow shop before moving toward JIT probably reflected that orientation in its MRP system, thus fewer radical changes would be necessary. The conversion of MRP systems is a case-by-case situation. The most severe systems rebuilding occurs if a company greatly changes both in physical process and in method of control; that is, from a highly transaction-oriented, detail-scheduled job shop approach to a minimal transaction, pull system, high-visibility flow shop.

If a pull system of control is used for a flow shop, final operations will draw the parts and materials they need when needed. However, supplying operations must be prepared to respond. They need a plan for each schedule period stating the daily parts quantities that will be required and the rate at which they will be used. From this, supervisors and operators can prepare layouts, tooling, and workplace organization to respond. Some deviations from exact quantities can be handled if all areas have prepared for that possibility in advance.

If material is to flow like clockwork, material handling and transport schedules must also be prepared so everyone knows the regular pick-up and delivery times. Once supervisors know those and the part volume estimates for the period, they can prepare detailed area plans: tooling, setups, maintenance, quality checks, personnel, and so on. They are close enough to the action to integrate the level of detail.

An overview of this planning is shown in Figure 7–2. Since few companies can put all parts on some version of pull system flow at once, their planning systems look a little like Figure 7–3 and might remain that way a long time. Part of it sets up pull system operations. Part of it does more conventional time-offsetting and gross-to-net calculations for long lead time and large lot-size parts. Those should be placed in an inventory from which a pull system can draw. The inventory must be controlled—visibly or otherwise.

Unfortunately, Figure 7–3 understates the magnitude of the potential mess. Some companies base their master materials schedule on a final assembly schedule, and their bills of material are structured on final assembled items. Others base their master schedule on forecasted use of families of material that are stored in inventory when finished. Final assembly later draws from that.

The bills of material for those two kinds of systems may be very different, and they may differ in other particulars. In such a case, a company may be well advised to just start a new MRP system and let it consume the old one as physical changes permit. Reform of the old one is too complicated. Running dual systems is messy, but in the end the company should have one simple system and less waste.

Inventory hidden in stockrooms must be accounted for. Each "random-time, random-quantity" put away and withdrawal must be recorded. Standard containers with standard numbers of parts in each simplify this procedure, but it is still work. Stock levels must be periodically verified and cost valued. In addition to stockrooms, inventory siphoned off into other sizable clusters not under strict visible-limit control needs some form of transactional verification—regardless of nobility of purpose: quality testing, traceability, special allocation, etc.

Every time a part number is set aside in a significant little pile, a system has to record it. Figure 7–4 diagrams a "conventional" production process. It is a simplified sketch of lead times in several categories for only one case of manufacturing. The same plant often has several cases. Each activity shown in lead time has some kind of system associated with it. All activities and delays shown might be separate classifications in a complex materials system.

FIGURE 7–2 Materials Planning, Pull System Parts

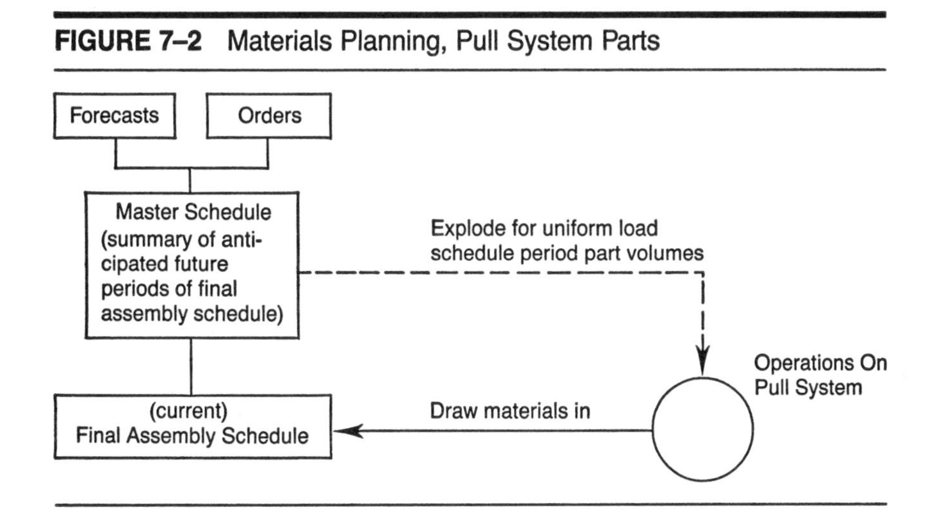

Every time an activity disappears, the system should withdraw from it and contract, but in many companies systems are huge and expensive. An effort can bog down because analyst and programming power is not sufficient to cut through all this muck. Managers, once dependent on systems as their handle on the business, are fearful of the chaos that would result from disturbing too much at once.

Figure 7–5 is a simplified sketch showing how a hierarchical bill of materials structure could modify and simplify as actual processes change. This view is indeed simplified because if bills of material used for materials planning are not based on end items, the entire set of planning bills may need to change. In addition, most companies with MRP have more than one bill of materials. Some have four or five, each with its own systems purpose.

Stagnation in these problems is the major reason JIT champions must keep everyone's attention on slicing out unnecessary activities and bridging over old systems. Otherwise, no one will ever wade out of complex materials (and other) systems.

The particulars depend much on the individual case, but a few common materials systems conversion problems are:

Inventory Accounting. As long as inventory is kept in stockrooms, it must be safeguarded by a transaction system. These trans-

FIGURE 7–3 Materials Planning, Mixed System

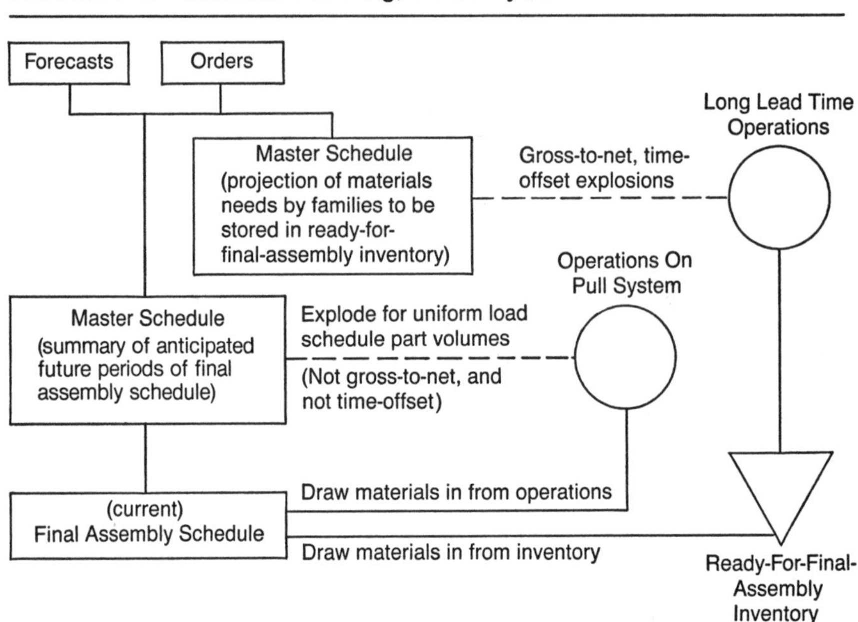

actions can be eliminated when inventory is in standard containers in visible control, as in a pull system *under disciplined limits*. Attaining this discipline is more than a one-week conversion; thus, for a time, production controllers and accountants will sweat out count verifications daily on the shop floor.

A small amount of inventory on a shop floor can be quickly counted, more so if it is in designated locations and standard containers. An hour or two once a month should do it. Counts are taken by the operators. Supervisory efforts to strictly limit the amount of inventory should result in a shortage being much more serious for reasons other than a data error.

Eliminating Inventory Transactions. If a company wishes to terminate processing inventory transactions every time material is used, it can do so by post-deduct or "backflush." Multiply the number of assemblies complete by the number of parts in the bill of materials for each. Subtract these numbers from the work-in-

FIGURE 7–4 "Typical" Manufacturing Lead Time Summary (make-to-stock, single-stage distribution)

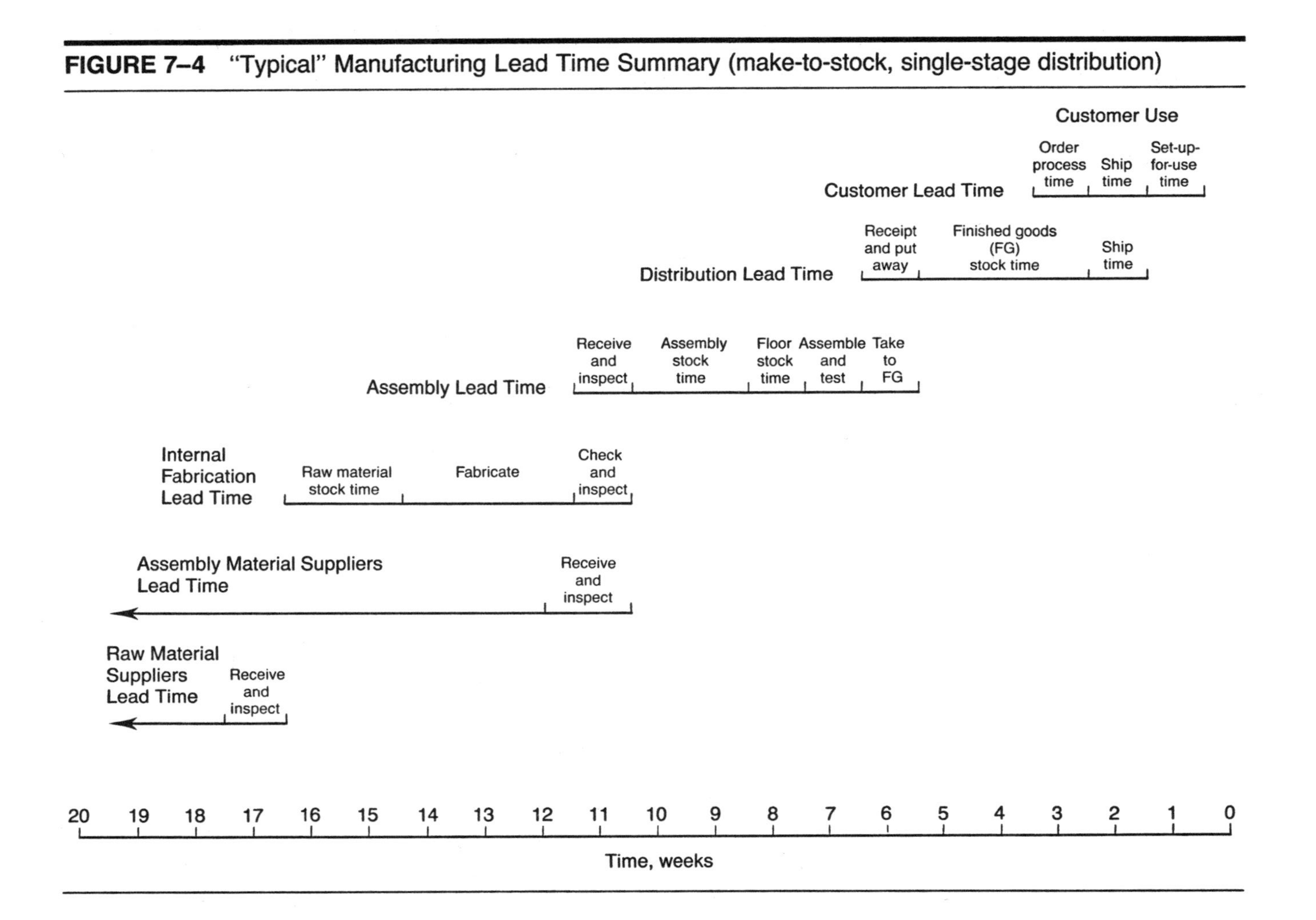

FIGURE 7–5 Structural Changes in Planning Bills of Material (assuming master schedule based on final assembly)

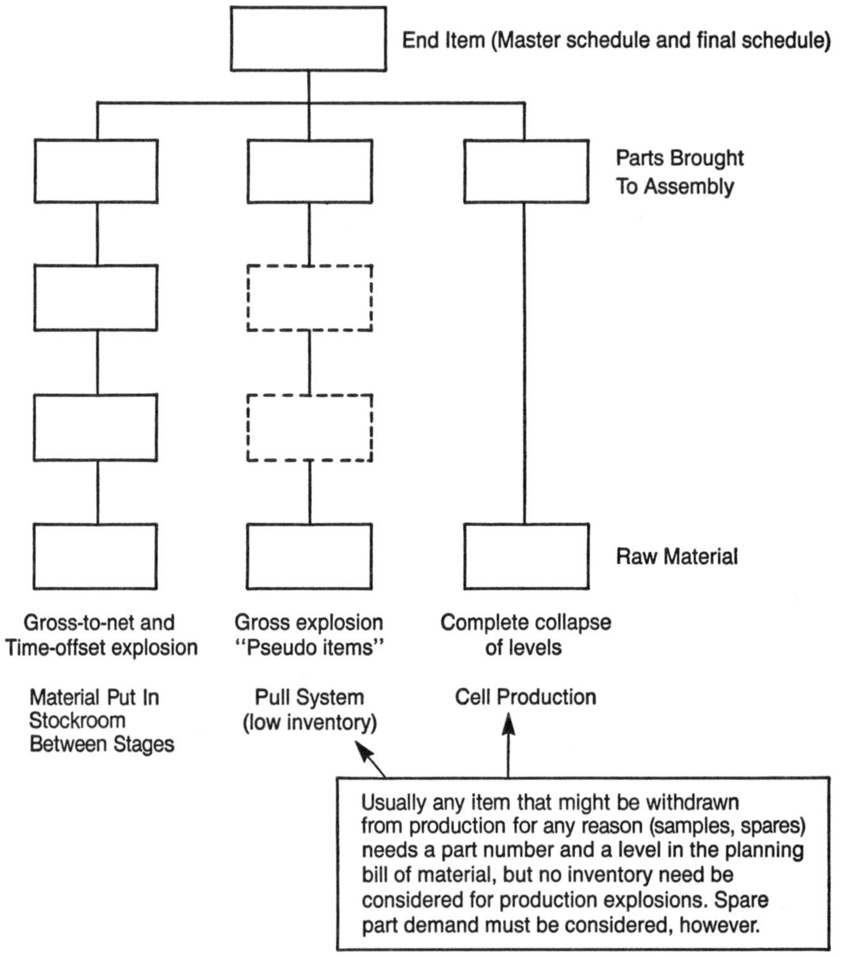

End Item (Master schedule and final schedule)

Parts Brought To Assembly

Raw Material

Gross-to-net and Time-offset explosion

Material Put In Stockroom Between Stages

Gross explosion "Pseudo items"

Pull System (low inventory)

Complete collapse of levels

Cell Production

Usually any item that might be withdrawn from production for any reason (samples, spares) needs a part number and a level in the planning bill of material, but no inventory need be considered for production explosions. Spare part demand must be considered, however.

process inventory figures. A company with visible, straight-through production should be able to maintain inventory just by counting parts into process and backflushing them out.

The problem is that if the MRP system or inventory systems are not programmed to do this, reprogramming is not easily done. Most companies develop a dummy transaction record from the backflush process, then play the record against the system, fooling it into thinking transactions took place.

Reprogramming eventually has to be done, but save up the changes for one trip to major surgery rather than do the job many times as changes take place a few at a time.

"Turning off" Modules of a System. The problem is that doing this also turns off data processing still wanted for other purposes. Suppose a quality hold area is eliminated, but the system will not run without verification of quality hold counts. Set counts to zero and keep them there, or perform whatever other dummy data entry is necessary.

The hazard is destroying some form of data feed needed for the system to perform replanning. Systems are minefields full of such things, and the only advice is caution.

A major process revision and simplification eventually requires a new system. The instinct of systems people is to search for it too soon—before they are fully into a new method of operation. They spend so much time doctoring old systems that they seek relief, and getting sidetracked into systems churning is a danger.

To break out of this, concentrate effort on attacking the physical process, eliminating unnecessary activities and delays, and determining how to synchronize the company into uniform load segments of schedule. New systems can be built around that, but not before the framework for them is in place. When the framework is in place, a serious push must be made to kill off an old complex system and get into a new one. A large group of people cannot live long with either a bandaged system or dual systems. The confusion is expensive, and sinking back into the morass is too easy.

COST ACCOUNTING SYSTEMS

To determine a cost, start with two questions: (1) Who wants to know the cost? and (2) Why do they want to know it? Costing has many reasons, some at cross-purposes. Some of the major ones are:

1. Cost-based pricing.
2. Operating decisions: make/buy, distribution changes, process changes, etc.

3. Performance measures of managers and units.
4. Financial reporting.

Ideally, one would like a simple, flexible cost system that provides numbers relevant to any purpose while still giving consistent results. This is self-contradictory and therefore impossible.

Assign a business case to a group of intelligent M.B.A.s, some of them practicing accountants. Ask them to cost the same proposals in a situation calling for judgment determining relevant costs and allocating indirect costs. Results will vary considerably depending on student assumptions and methods.

There is no reason to expect universal agreement on cost models. Many accountants might agree that one is especially bad for its purpose, but hardly any will agree on the one ideal method to cost the same situation. All accountants should arrive at the same figure for a cash account and for the valuation of payables, receivables, and inventory—though even these require judgments on collectability, obsolescence, and the like. Cost allocations and attributions are judgment calls.

If everyone in a company had a personal computer and organized costs as they deemed reasonable to fit the situation, there would be no agreement among them without arguing out their assumptions and methods. Companies try to develop consistency by adopting "official" costing procedures and building systems around them. However, all cost systems have assumptions built into them: systemic cost biases.

Cost figures must be compiled in some way, and the systemic biases come with them. However, decisions do not have to be based on cost figures known to grossly mis-model alternatives considered. Fully loaded cost figures used for price basing may not be at all appropriate for comparing operations alternatives. Managers and accountants may use them anyway rather than work through the issues of judgment involved, and thus systemic cost bias affects the direction of management decisions.

Some of the accounting issues associated with JIT/TQ are:

1. Interpretation of costs for management direction.
2. Systemic cost biases inappropriate to future management direction.

3. Complex cost systems.
4. Changing cost systems to better fit a new process.

Interpretation of Costs for Management Direction

Manufacturing excellence probably cannot be attained by a management whose judgment is utterly cost dependent. Progress is seen as a series of independent cost decisions rather than as a long-term direction, and judgment never rises much beyond the systemic biases of the cost methods used, whatever they are.

This amounts to a rather single-minded management pursuit of provable cost reduction. Morale improvement, visibility improvement, or even cost-benefit of defect elimination are extremely difficult to prove out by any system of costing that considers only tangible relevant costs. Such a management will reject the notion of spending $15,000 to change a layout that will save $15,000 in floor space on grounds that nothing is changed, and the space will not be used for anything else.

The same management might well spend $15,000 for extra space at a marketing exposition. No one can finger any tangible results in advance, but without marketing presence over a period of time, market position will slip, and so continuing effort must be made. Some of it results in a payoff.

The same is true of continuous improvement effort in manufacturing. Some changes will undoubtedly be mistakes, and cost information is important because no one wishes to go broke with one mistake. By the same token, costing should not exercise veto power over every move.

The perception of production as a collection of independent cost centers is a problem for interpreting costs also, no matter how the independent costs are collected and structured. The assumption is that driving the direct costs of each of 10 operations to a minimum will reduce total cost. The vague hope is that total overhead and indirect cost will reduce at least in proportion.

However, indirect costs and overhead are reduced when the total pattern of production is simplified. When production is complicated, the costs of managing the whole mess must increase, which comes out as indirect and overhead cost.

An operation working to cycle times has an excellent basis for costing in the cycle times themselves. Planned cycle times, multiplied by cost factors, are planned costs. If production is always caught up with the schedule at the end of each shift, the only variance from the cycle time cost is any necessary overtime. Detailed operations variances from costs not based on cycle times mean little.

A company zeroing in on defects keeps good records on defects. Defect reporting is the basis of scrap costing, and there is no acceptable defect level, but rather targets for improvement.

As a company develops its thinking into continuous improvement, it begins to question the value of standards which may have extra time and "normal" defect allowances built in. If the value of the standards is suspect, there is little reason for reporting variances against them. The reporting of actual costs based on cycle times and the actual data from operations is relatively simple. A production operation is likely to have two costs: actual costs and target costs based on the various targets for improvement.

Before launching a major campaign to revise cost systems, first examine management need for cost data and management attitude toward managing with costs. Two extreme examples may help.

A plant controller was appointed plant manager and immediately decreed that from now on the plant would be operated strictly by budget. Detailed budgets were already established operation-by-operation, and the man poured over the variance reports daily. Managers and supervisors were penalized for being either over or *under* variance. So much for major cost reduction. Any such projects were exceedingly careful and cautious so as to be manipulated within cost budgets. Overall direction: Zero.

At the opposite extreme was a fast-growth company with undeveloped process costing. Product costs were estimated according to which of three classes they belonged, determined by how many of over 200 operations they passed through. It was impossible to know for certain whether the total production cost of any single product had moved plus/minus 10 percent, though a guesstimate could be derived from the changes in routing and yields before and after a process change. The company did not even have an aggregate beginning benchmark cost by product.

The moral: Try to set some initial benchmark costs by plant, by department, and by product, but do not delay everything while waiting on this. Better to have cost reductions than to know precisely how much they were. Track the reductions in overall costs and keep tight reign on overall budgets. Let managers use judgment about detailed changes so as to bring them together into an overall pattern that will produce an overall result. Accounting issues begin to clarify with an overall direction toward improvement.

Systemic Cost Biases

The most common is allocation of high overhead percentages to direct labor costs. As direct labor becomes a smaller and smaller percentage of total cost, allocations become distorted; and by the time direct labor is 5 percent or less of total cost, almost everyone thinks the costing is crazy.

Overhead costs may go through several stages of reallocation among each other, but in a factory they all come to rest on direct production costs, allocated on some basis. Other than direct labor, a common basis is machine time. Companies transitioning to JIT/TQ talk about allocating cost in other ways.

One method is allocating on the basis of material throughput time or by level of work-in-process inventory. This is presumed to indicate the consumption of overhead resources. The problem with it is the necessity for tracking the base measure (throughput time or WIP), which is not difficult if production is uniform loaded. However, interruptive rush work that occasionally enters such a flow needs to be considered. If it receives almost no overhead load, its cost will be "unfairly" low. Those jobs really take the overhead in such a system, and they will appear to be more profitable then they really are.

Another method is allocating cost on the basis of transaction volume required to process each product through a plant. These are countable if defined as such and if a computer system is programmed to keep them. This method promotes the introduction of low-transaction, simple processes but also the suppression of transactions, which might not be the thing to do universally.

Why not allocate costs on any reasonable basis for cost-based pricing and financial accounting, but make operating decisions based only on direct, relevant costs? This old controversy usually bogs down in the necessity to estimate some costs by allocation for operating decisions. For instance, what is the cost of maintaining a machine?

Direct costing in the pure sense probably cannot be adhered to for operating decisions, but this direction seems sensible. Keep cost systems simple. The purpose of costing for operations decisions is to avoid the really dumb mistakes.

Complex Cost Systems

Not every manufacturer has a complex cost system, but many do. Ask most managers which systems chew up the most computer time, and they will say materials systems and accounting systems. Conceptually changing cost systems seems easy, but changing a complex one is subject to the same problems as changing a complex materials system.

Complex manufacturing accounting systems often have their own bills of material similar to, but slightly different from, the bills of materials used for materials systems. They are used for denoting the points where costs are accumulated in the production process. If a design change or process change occurs, the accounting bill of materials—and the cost structure it represents—also changes.

Accounting systems are no more programmed for versatility than most large materials systems. They do not ease the transition into JIT/TQ thinking.

Changing Cost Systems

Probably the worst-case scenario is conversion from a job shop operating method with job orders to flow shop methods without job orders. The job orders represent the transaction medium to which most cost reporting is attached. If they are no longer present, the method of costing must change from job costing to some form of process costing. Both are textbook costing methods, but conversion from one to the other is both work and trauma.

The trauma occurs because the data collection of costs for financial accounting is affected, which is more serious than cost reallocations for managerial accounting. This brings up the problem of reconciling costs under a new system with those under the old. Otherwise, no one knows whether a reported profit (or loss) is "real" or just a figment of restating the basis of costs. Auditors (and tax authorities) insist that financial statements be consistently stated from one year to the next, thus systems must operate in dual long enough to benchmark comparisons.

In addition to this aspect of system change, no controller wants to be in a position where costs seem unaccounted for in internal reporting. A conscientious accountant is likely to spend many hours verifying that visible control systems really work and assets did not disappear somewhere.

VISIBILITY OFF THE SHOP FLOOR AS WELL AS ON

If all this system changing seems a bit complex, the principle of simplifying it is visibility. Try to avoid the situation in which each analyst off the floor sees only a small part of the system dealt with. Post schedules in offices as well as in factories. Post targeted objectives and attainment toward them. Try to make systems do double duty. For instance, try to make the accounting system and the materials system use the same bills of material. Make transactions serve multiple purposes. For instance, one scrap reporting transaction should serve for accounting system, materials system, and quality system.

Staff instinct is to crawl off in a corner with their own little pile of data. Stimulate interaction by having them use the same data. (One set of data from one source is a staff version of workplace organization. Keeping data at the scene of action helps too. Those gathering it know what it is. Those seeing it can easily put it in context, and not too much will be asked for.)

It seems inefficient from the viewpoint of one special staff job, but the objective is to keep a high level of total awareness so that time is not spent correcting mismatched data, correcting errors, and discovering other staff's oversights. JIT thinking applies to support work as well as to shop work.

Finally, any system is foremost a *people system*. The human is an integral part. Integrative systems design is accomplished by developing people and systems together.

CHAPTER EIGHT

The External Factory—Suppliers

Half of the average manufacturer's production cost is purchased material. If internal production is mostly assembly, as much as 80 percent of cost of goods sold is pass-through from suppliers. Competitive manufacturers must manage this external factory by the same philosophy as the internal one.

Think of a materials supplier in the operating sense. The true supplier is the last operation touching material in a remote facility. The true customer is the first operation to use the material received. Anything that unnecessarily complicates a simple hand-off of material from one to the other is superfluous—waste.

Other activities are necessary to make the actual material transfer smooth and simple: product design, certainty of quality, schedule, containerization, transport, and financing. If one is trying to eliminate waste, detailed decisions on all these should not be necessary for each shipment.

This simple concept mightily contradicts established business practice. American companies find it difficult to strip away the layers of custom that smother operation-to-operation supplier relationships with different ownership at each end of the line. Doing so is not easy if the supply relationships are inside the same company.

In large companies, a frequent complaint is that the worst suppliers are other plants belonging to the same company. Speculative reasons for this begin with the observation that "internal" suppliers seldom have a marketing department campaigning for internal customers' interests with the same vigor as for external customers. A sale with exchange of real money wins more attention than one checked off with a mere accounting transfer.

Companies try to counteract the apathy by having internal suppliers compete with external ones for business. A little excitement is added along with increased negotiation expense, but the result does not necessarily approach excellence. To attain that, managements must become excited about the dull subjects of quality and waste.

In Figure 8–1, a single rolling mill should feed coil stock to the same press. Why? Because one problem in quick setups of presses (and attaining consistently good parts) is nonhomogenous coil stock. Gauge variance and inclusions must be compensated for.

The same rolling mill is a consistent physical source. The conditions of roll forming should be known, and the mill should be interested in improving the quality of output for its users. If it does not, go to the *physical source* and work it out immediately, just as if the operation were part of the press shop production.

This action sounds logical, but small companies do not buy from a mill. They buy from warehouses, and the mill source may or may not be consistent. If the steel is inadequate, the expedient recourse is to shop for better.

Even for large companies buying mill-direct, having press shop technicians and operators visit the rolling mill and directly discuss particulars with mill technicians is not an idea many would take seriously. Mill technicians have time to talk to only a handful of thousands of actual users of their output, and a squad of various technical and managerial intermediaries would have to be straight-armed for the actual users to get through to them.

This same situation is all too often true of castings from foundries, motors from winding operations, and other materials coming from production processes separated by geographical distance and great organizational chasms. Barriers to operation-to-operation production networks are geographic, traditional, organizational, and perhaps even legal in some instances.

The objective of an operation-to-operation network of suppliers is constant improvement, and within that the first priority is quality. The external factory is a mutual improvement society. It should be organized so that if something goes amiss, the concerned party should know who to contact at the source of the problem and how to go about making improvement immediately. Quality service from a supplier is more than having an SQC program and prompt delivery—a techniquish view of JIT/TQ. The causes of problems do not

FIGURE 8–1 Operation-to-Operation Suppliers: Suppliers as a Production Network

Think of suppliers as specific production processes that are linked to your own, but from distant locations:

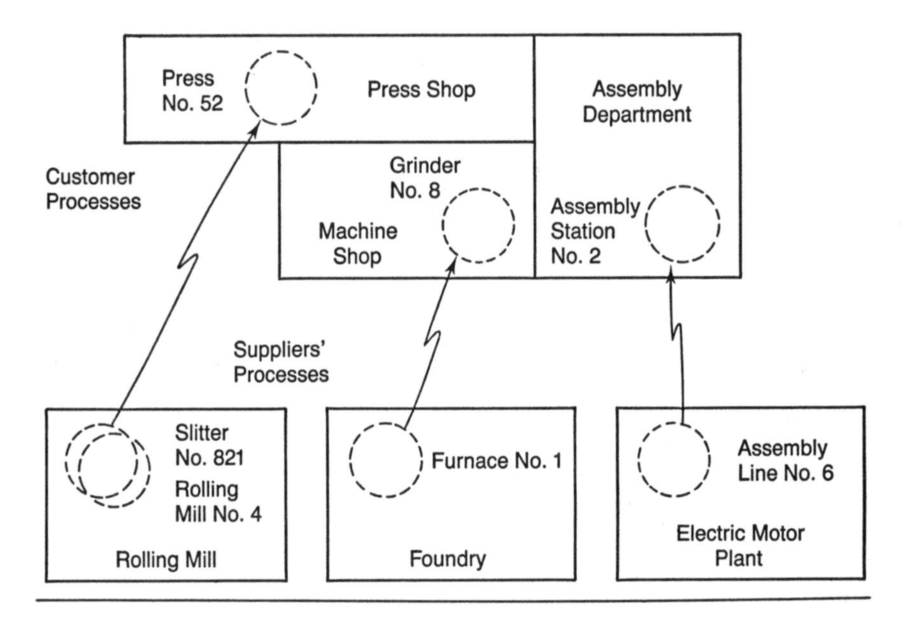

always reside with the supplier companies either, so this is not a strictly dictate-to-the-supplier arrangement. The need is for organization-to-organization relationships to promote quality and eliminate waste. The techniques come from that.

Many U.S. companies have in recent years tightened quality requirements for suppliers, and many have arrangements for daily deliveries from at least a few suppliers. But most of these arrangements still do not approach the necessary spirit of the operation-to-operation network—something that pervades the relationship between two companies when it really exists. Most of the great examples are still Japanese.

An incident touring an American JIT/TQ plant with some JIT/TQ-oriented Japanese illustrates the point. The American plant had made great strides, but at one operation an irritating problem was observed with a purchased part. The instinct of the Japanese was to *immediately* go to the physical supplier operation and work it out. This did not occur to the Americans; for them, "immediately" was as soon as appropriate persons could organize to focus on the problem intelligently and figure out a solution. The Americans began by mentioning the proprieties of purchasing, marketing, and engineering involvement by the two companies. To the Japanese, much of this was time wasted and merely delayed getting to the core of the situation.

Many large American manufacturers have computer links directly to suppliers, or they are contemplating them. Usually these links are planning-system-to-planning-system, or they are links to track shipments. Too few are based on thinking through what is necessary for a minimum-waste linkage between two operations, with direct, immediate feedback for quality. (A few may be getting close. For instance, Lear-Siegler builds and delivers car seats to Chrysler on-line and real-time in assembly sequence.)

Companies must start internal improvement programs first. Those who start to develop suppliers should be far enough along themselves to understand what they are trying to achieve. Many companies push suppliers into carrying inventory and refining delivery methods before either operation is ready for it, sometimes adding to waste rather than subtracting from it.

Mutual trust is an obvious issue. Understandably, a supplier will not divulge its trade secrets in the interest of upgrading a customer's product quality if there is the slightest hint that the customer might compete in the same business or provide tips to competitors.

In other words, developing a shipping schedule and containerization program is only a part of mutually improving quality and reducing cost. The core assumptions of how one does business are shaken. Negotiating strategies change. Concepts of whether one got a "good deal" alter.

Ask the average business advisor whether a purchase agreement is a good one, and the instinct is to say yes if the work went to the lowest bidder qualified for the work. A set of long-term supply relationships is apt to be seen as a cozy arrangement to fatten the select few, and without competent people in both camps being pressed for continuous improvement, that condition may well develop. Judging the ongoing merits of an operation-to-operation relationship requires product and process knowledge. Anyone can check whether bids were let.

An investigation to determine the cause of either poor supplier quality or poor delivery is likely to lead back home. For one reason or another, the people actually doing the work for the supplier did not understand what was needed. Unclear changes in either specifications or schedules can be the cause. Supplier improvement is in part improvement of the customer company.

All suppliers need constant attention to improve quality, delivery, and cost. Great potential exists in long-term, close relationships with selected ones: joint process development, joint product development, mutual planning and control—almost as if one company. This intertwining of operations must be disciplined—keeping "tightly controlled inventory" operation to operation, whether physical operations in factories or planning operations (as between engineering departments). Otherwise, comfort will bring deterioration to operations between companies just as surely as inside them.

This kind of relationship can develop with only a few suppliers. One can work closely with only so many people. Rather than arm's-length negotiating between strangers, the relationship is based on technical and operations problem solving. The capabilities of people should develop in both companies. The change is somewhat like a major league baseball team deciding to reemphasize development through the farm system rather than buying talent outright.

SELECTING SUPPLIERS

The first phase is deciding those suppliers with whom a close working relationship might be beneficial. Perhaps one or two exist already. For the rest, begin by evaluating their purpose to the customer company, searching for long-term reasons why the companies should need each other. A full evaluation may not be possible until a company is far enough into total quality to understand its own requirements thoroughly:

1. *Commodity suppliers.* In some cases, fasteners may be commodities; in others, definitely not. It depends on whether a special design is necessary, and whether quality and consistency much

affect the quality of the final product without waste. If the general level of quality from any established source will serve interchangeably, the item is a commodity, and no special source is needed.

However, classification of an item as a commodity is not always obvious. As noted, coil steel may not really be a commodity although current purchasing practice might treat it so.

2. *Technology suppliers.* In addition to patent licensors and contract R&D firms, some suppliers of material are of prime importance because they are innovators, or potentially so. Which manufacturer of integrated circuits will first be able to provide *your* design in surface-mounted form and produced in volume is problematic. Sole-sourcing a loser is not the way to go.

Primary suppliers of capital equipment—machines, tools, and instruments—are in this category. They are sources of process knowledge. Close work with them is necessary to refine process capabilities and modify equipment for flexibility and fast setups. (The condition of the machine tool industry reflects not only on itself but on its customer base.)

3. *Processing economics.* A common example is plating, a messy process requiring both skill and capital and fraught with hard-to-manage environmental concerns. Many companies who could operate plating if desired will choose not to for several reasons:

a. The volume of plating is low and the fixed expense of people and equipment high, and they do not wish to be in the plating business for other people—the classic case.

b. Their location is such that contending with fumes and acids is not feasible.

c. Wage rates for plating are too high or too low to be compatible with the rest of the plant. Utility costs are lower elsewhere, or other factors strongly favor a distant location.

Prime candidates for organization-to-organization and operation-to-operation relationships are in categories 2 and 3. They provide something unique to the customer company and thus have reason to be partners in matters of product design and production process.

Other things being equal, a "JIT supplier" geographically closer is better than one far away. Personal contact for joint work is easier, which is a separate consideration from just transport distance and frequency. However, proximity is not the only consideration. Better a quality supplier at a distance than a mediocre, inconsistent one close by.

Take care courting suppliers whose parts could foreseeably be produced in-house as a result of internal improvements. Little is more damaging to supplier relations than promise of a long friendship followed by a doublecross; but suppliers aware that they are candidates for omission will usually generate great interest in how they can improve. If they do, try to clarify their status as soon as possible. If they do not, the decision is easy.

Emphasize quality of process in supplier selection. Delivery is important as an adjunct to improvement of quality. Stress immediate feedback and quality improvement. Suppliers who will open up their data for mutual problem solving on that basis are good candidates for further development.

Evaluate the people and their willingness to dig into the source of process and quality problems. Mere demonstration of control chart capability is no assurance that quality will ever approach parts per million. Promises by sales personnel are insufficient. The companies must form a bond between engineering groups, scheduling staffs, traffic managers, and line management—operations-to-operations working contacts.

Many JIT companies early in their odyssey have held supplier days at whatever plant shows enough progress to provide a demonstration. These are very useful but not the place to pass around the sign-up sheet. Few company representatives attending these affairs can commit their companies to anything other than modest effort.

At these meetings, learn who else in the company should be contacted to forge operating relationships. If the supplier holds promise of developing, evaluate it closely. Go there, meet them, and study supplier operations firsthand. Get to know some of the people, including operators. Going for a long-term relationship is a lengthy, expensive courtship, so target the companies for this attention carefully. With other suppliers, as the saying goes, "Do good every chance you get."

Some of the early work is discontinuing contact with occasional suppliers used only if a severe problem arises. An objective is to improve and stabilize until no one need call a rarely used supplier to bail them out. Just cutting the total supplier list by half or two thirds simplifies purchasing management.

Most companies making great internal improvement are dismayed by the sluggishness of suppliers. At least one major obstacle to internal improvement is likely to be a supplier who cannot accept that performance once tolerated can be no longer. No simple solution is available. The only short-term alternative is to bury the problem in inventory—sharply defined, red-flagged inventory. A quality supplier must be developed long-term.

Keep the operation-to-operation concept in mind and build on it. Look for suppliers whose operations would fit well into a stable, uniform load schedule if one could be developed. If a supplier can produce a family of customer parts from one operation in one of its plants, the overall stability of schedule to that operation can be improved. If one part is not needed, another is, and so the operation keeps going. Evaluate the potential of suppliers' operations for this as one might evaluate the development of departments in one's own plant.

Look to develop a supplier network that will decrease total volume of interaction for routine matters. A transport pattern among physical locations should be put on as regular a basis as possible. Without first seeing to operational improvements by both parties and tending to the nature of the schedule presented, pressing suppliers for refinements in delivery is asking for results before development.

Larger companies will find a form of regular contact with suppliers useful, perhaps even a supplier organization with regular meetings. Invite the smaller ones to participate in training that they might not be able to afford otherwise. Inform them of new developments without waiting on a marketing representative to make communication. Invite their engineers to participate in new designs. Arrange for their supervisors and operators to meet those who use the material they send—their real customers.

Such an approach runs counter to many existing practices. For instance, the marketing strategy of many prospective suppliers is to work with company design engineers in order to be "designed into the business." Design policies of the company must be known to engineers and explained to designers. They should understand the need to look at a total supplier process; thus, presenting a sample part is only the beginning.

Xerox began to redeploy its supplier base in 1981 when its materials management costs (purchasing, materials handling, systems) reached 9 percent of the total materials budget. It taught suppliers SPC while tightening quality and delivery requirements. Buyer-engineer teams reviewed component designs with suppliers, and many value-engineered improvements came from supplier suggestions—designing to the suppliers' processes. The total number of suppliers selected for a new product line in 1985 was 300, down from 5,000 in a comparable situation in 1981. Xerox found the best suppliers, shifted the business to them, and dropped the rest. By late 1985, production costs had declined 48 percent and materials management costs were only 3.3 percent of the materials budget.

This activity is only sketchily beginning in the United States. Attending to suppliers is burdensome and very confusing if one plunges in too deeply while still trying to make basic internal improvement. Care and feeding of suppliers is one of Harley-Davidson's most vexing management tasks, for instance. Do not expect too much too soon.

SOLE-SOURCE SUPPLIERS

The terms "sole-source" and "single-source" are not always used consistently when referring to suppliers, but usually sole-source supplier designates the only one capable of meeting requirements for a material—usually technical requirements. A single-source supplier currently has 100 percent of the business for a part, but if they fail to perform an alternative source is available. Although the alternative source may not respond as quickly as the single-source supplier, their response would be faster than if another company were to be developed to meet technical requirements.

A single source in a supplier network is often considered a single shipping point for a particular part or family of parts. For example, in automotive assembly, a common objective is to have as many single-sources as possible shipping into a single auto assembly plant. This simplifies the problems of managing a transport network into the plant. However, a large company with multiple assembly plants may have different suppliers for the same part, each of whom is a single source to one or more assembly plants.

The risks of sole-source and single-source suppliers are well known; the benefits of developing these arrangements are usually underestimated. If a company develops a sole source for a uniquely specified material, the specification must be conveyed and the results verified only once. Multiple sources are multiple development projects requiring more time and work. Even then the same part from two different production processes in the same organization may both meet specification but not have identical characteristics.

Should a quality problem arise, tracing the cause is more difficult with multiple sources. With a single source, traceability is simpler; and, with immediate feedback, responsibility is easier to establish.

In JIT/TQ, the notion of a single point of production feeding a single point of use is basic to the integrity of a network of quality responsibility. The driving motivation is creating a condition in which problems are openly addressed, waste eliminated, and quality sustained. Go for 100 percent reliable quality performance.

Companies fear that a single source will have a fire in the plant or a strike. However, a sole-source supplying organization may have multiple operations or multiple plants. Ordinarily one process feeds only an established set of customer operations, but in an emergency, some reserve capacity in supplier operations dedicated to other networks can be called to cover it. One localized calamity does not destroy everything, and if the supplier has good people involvement, strikes are minimal risk.

Another reason for linking suppliers operation-to-operation in single-source arrangements is schedule stability. If part use by a customer operation is constant, but multiple sources supply in varying percentages, the demand pattern each sees is less stable than the overall part use from which they are derived. (Averages over time do not count.)

Managers fear that a sole-source supplying organization will press its leverage. A few have. The idea of open problem solving was lost on a new incumbent of management at the supplier end. A stiff price increase seemed a good economic signal to terminate an entangling arrangement he did not understand.

However, sole-source suppliers are not new, and many have endured for years without price gouging and lackadaisical delivery. The bottom line in these relationships is *people* and an appreciation that suppliers and their manufacturing customers are not really in competition for each other. The real competition is for the customer who uses the end item produced.

Both supplier and industrial customer have mutual interest in improving quality, delivery, flexibility and cost (or at least they should have). However, negotiations between them can poison this mutual interest if the parties misperceive that a highly competitive situation should exist between them. In practice, this cooperation is motivated by all parties realizing that an entire industry is in tough competition. No one is likely to be the beneficiary of a windfall for long.

If both supplier and customer are really into open problem solving and continuous improvement, that becomes the core issue of negotiations. If savings can be had, how should they be split between the parties? Something like a 50/50 split is the usual outcome, but that depends on whether there is any real surplus to be had by anyone. In a tough market, all the savings will go to the end customer, leaving nothing to split between the providers.

Where the problem solving is open, so is the negotiating. Both customer and supplier know each other's costs, so the possibilities for secretly padding the numbers used for negotiating are limited. The question of fairness begins to revolve around margins. The customer company would naturally like the suppliers to help keep its gross margin reasonable in a competitive situation, and a customer refusing to share a fat margin with suppliers tends to decrease the morale of suppliers running a tight operation.

The people managing a long-term supplier relationship have to think long term on both sides. These arrangements are not to be entered lightly and are therefore not broken easily. How long is long term? Long enough that the supplier and the customer can afford to invest capital and people in long-term development of each other—not forever, but longer than annual contracts. Product family lifetimes is a minimum time, and those involved should establish the relationship so that it will outlive the tenure of one or two people in their current positions.

Reality is that good relationships can go sour. Conditions and people change. With a sole-source supplier (or with any on whom there is long-term dependence), a good idea is for the parties to agree on the conditions by which the arrangement might be terminated. This is little different from the clauses in long-term leases of commercial property in which discontinuation by either side should be spoken far in advance, although voluminous legal agreements seem not in keeping with the spirit of things.

Long-term sole-source and single-source supply agreements are long-term problem-solving partnerships. The negotiations should be conducted as if hiring an architect or an advertising agency. Long-term suppliers are really in the service business. The service is providing quality improvement, waste elimination, flexibility to respond and so on as supplier and customer mutually develop to do it. The relationship between the companies must be built with people-to-people bridges at several points—line management, engineering, quality organization, and others. Unfortunately it is a long developmental journey to this state of thinking if companies are just emerging from order-at-a-time haggling and expediting.

SUPPLIER QUALITY

This is the beginning point with all suppliers. Working with supplier quality is an extension of improving internal quality. The objective is to eliminate defects coming from suppliers, and it begins by understanding what a defect is, followed by suppliers understanding that no defect is truly acceptable.

Many suppliers cannot understand it. Their first reaction is to ask for a price adjustment to compensate for doing what has previously been the customer's work of sorting out the defects. So begins a tedious process cultivating supplier quality.

Educating the Buyers

Suppliers are confused if the buyers who contact them are confused. Confused buyers may ask supplier salespeople to please send a copy of a control chart with each shipment, which does not particularly assure defect-free arrival, and the message will surely be garbled as the salesperson transmits it to the operations managers.

The role of the buyer must change drastically. Buyers must understand the operations of their own companies well enough to comprehend the issues of quality with suppliers. Seldom can a buyer do much to improve a supplier's operations unassisted. The role changes to building and managing a relationship in which a cast of characters is involved on both sides.

This skill is considerably more than jousting with salespeople, haggling price and delivery, and expediting as needed. It cannot be done well with a large number of suppliers. Furthermore, in the building stage, the buyer must deal with suppliers in all stages of quality development. The building stage may continue indefinitely.

Purchasing managers should not push inventory on suppliers for show, not demand that each supplier conform exactly to the customer company's statistical methods, and not try to assure quality by overspecifying parts. They are key figures in guiding suppliers to provide the quality needed as the customer itself comes to understand what it really needs. They cannot set their heads on that after only a day or so reviewing total quality.

Purchasing should develop quality goals for suppliers just as quality goals are set internally. They should be comfortable enough with total quality to formulate and execute a realistic supplier quality development plan.

Rating Suppliers

Companies using formal supplier rating systems usually base them on price, delivery, and quality, sometimes adding other factors such as flexibility and responding to changes. Such ratings assist in selecting suppliers from among a large supplier base but are seldom used for more than exhortation to the supplier community. Ratings may not even be disclosed to suppliers.

Supplier ratings mean little anyway when buyer performance is measured on price, price, and price. Then, supplier relations are trapped in haggle-and-expedite mode. Buyers should be rated on their work *improving total supplier performance* over time. Measure, target improvement, and track just for an internal operation.

Start by removing buyers' performance measures that inhibit true supplier development. Then, begin to rate suppliers as is necessary but usually emphasizing quality first.

Large companies use quality rating systems for suppliers that are generally modifications of systems used internally. Rating numbers or indexes are assigned by reviewing a supplier's operations. By delving into operational specifics, these have more potential for suggesting improvement, but they do not assure motivation for it. In fact, if the better performers receive some recognition as a "Q1 supplier" or whatever, care must be taken that the rating does not lead to self-satisfaction. The real criterion is *continuous improvement*—never being satisfied—but attitudes and understanding must be developed before this can be accepted.

A supplier interested in improvement must know what it should do. This comes first. Know what is important to the customer in specifications, presentation, and delivery. Work out the needs for supplier performance according to the status of that supplier's understanding and development. That is the beginning of "measure, target, and track," and both companies should understand that the source of some problems will lie in *customer* operations.

A rating system may provide guidance, but it does not convey the specifics needed to tell a supplier that, for instance, the customer would like to narrow the variance in capacitors received from ±7 percent to ±5 percent with no increase in price over the next year, if the supplier could possibly improve its process enough to provide it. For companies dedicated to continuous improvement, progress toward such a goal becomes part of the rating.

Including detailed improvement targets as a routine part of supplier negotiations is not advisable until the customer is far enough into total quality to work with the supplier in detail, and until the supplier has progressed to where it understands how to respond. Otherwise, these are seen only as unreasonable demands, figments of an imaginary future.

Working with Suppliers

This little phrase covers the heart of supplier quality development. The first issue is to be certain suppliers understand the standards and quality specifications necessary, which implies that first the customer company should understand them. All of the methodology applies, going back to understanding the requirements of the end product in use.

Just as is true internally, improvement of quality from suppliers is a matter of detail-by-detail work. The customer company will not know the suppliers' processes in the same detail as its own, but the approach is the same. Progress comes from finding and developing suppliers who will work this way.

Detailed work with suppliers cannot take place rapidly. Time is not available for it. The select, primary suppliers need regular visits, which consume time. How often should they occur? No general answer applies, but as the plant manager of one supplier company observed after a customer visit, "We found firsthand what we needed to clear up. We really should do this more often than once every 10 years."

Pick suppliers you can work with in detail as part of the selection process. Those may not turn out to be the ones highly rated at first. If they are initially bad in price, quality and delivery but are capable of improvement and willing to try, they may turn out the best in the end. Some who start in good positions remain stiff-necked about it. Suppliers are people, too, and therefore frustrating.

Some suppliers may be ahead of the customer. Some will be eager but not understand. Others never comprehend why they should understand, as with the owner of a small business who declared, "I don't tell my customers how to run their business, and I don't expect them to tell me how to run mine." (The company makes a unique product but has serious quality problems because it is "difficult to work with.")

Finally, urging quality on suppliers is not attained by imposing contractual requirements to use specific quality control methods. The methods may only be eyewash for the customer, not appropriate to the supplier's situation, and, if they are appropriate, not used with understanding. In addition, a supplier having such demands from 10 different customers cannot possibly comply with all of them without turning its own quality management into wasteful bureaucracy.

MATERIAL TRANSPORT

The objective is to reduce the *total* cost of transferring material from the supplier's last operation to the customer's first operation. The transport of material between plants may be only one of many steps in this process. The idea is to eliminate as much waste as possible by as direct a hand-off as possible between the two operations. That may demand attention to many factors other than just the method of transport, as shown in Figure 8–2. The flow diagrams omit detail to illustrate the point.

A quick impression of JIT delivery is closely timed delivery in small quantities at frequent intervals, which is true as far as it goes, but to prepare for operation-to-operation transfer, a good deal of silliness can be introduced. Adding to transport expense without a payback through less waste in the total process makes no sense.

One major consideration is quality—confidence that material can transfer without any special stops for inspection at either end of the line. Another is schedule. If the operations are not timed to

FIGURE 8–2 Objective of JIT Transport

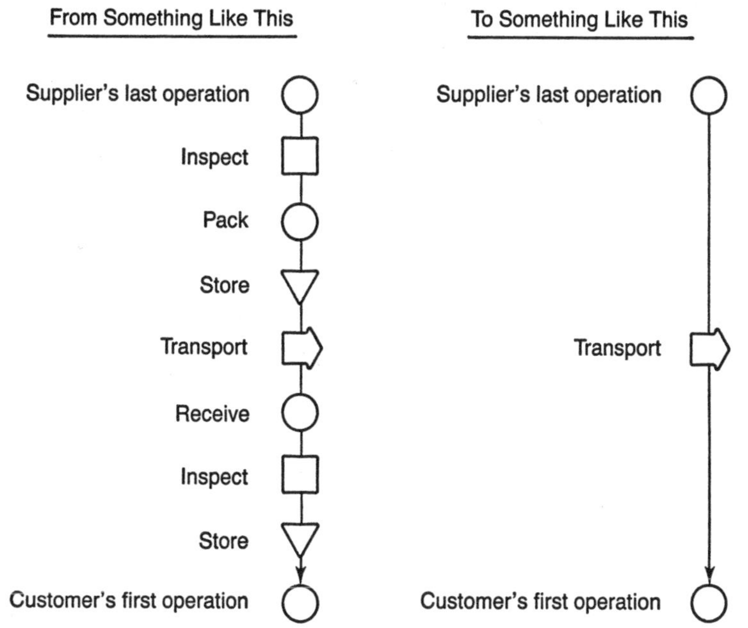

From Something Like This

Supplier's last operation
Inspect
Pack
Store
Transport
Receive
Inspect
Store
Customer's first operation

To Something Like This

Supplier's last operation
Transport
Customer's first operation

mesh together, material must be stored somewhere to wait out the delay.

Much of JIT delivery's mystery is removed by thinking of it as developing the total process of the two operations at either end of the line and the transfer process between them. A sharp, closely timed hand-off between operations is not attained without working up to it, but lead time (and total amount of stock) between the two operations can be cut well before perfect quality and perfect synchronization is attained. Follow the principle of immediate feedback to improve operations.

However, improving the timing of this performance requires holding a stable schedule while supplier operations are actually in process. Where possible, a uniform load schedule provides a repetitive pattern for synchronizing suppliers just as is true of internal operations; and, if that can be held, potential for close timing increases. Not holding it results in some form of increased delay between supplier operations and the customer, and schedule changes cause expediting—unplanned, chaotic JIT. One major reason for not holding schedule is supplier nonperformance.

This can appear to be a permanent trap. If a supplier—one of many—does not deliver or if its material is unusable, either the customer process stops completely or a schedule change goes out to many other suppliers to try to keep everything running. In this chicken-and-egg situation, the uniform load schedule comes first. Without it, there is no pattern to work against, however poor the actual performance to schedule may be. The rest is supplier development—working with them and working to improve internal operations to avoid schedule interruptions.

This problem, like many others, has no pat answer. The principle is to chip away at the product design, the quality issues, the transport problems, and whatever can be isolated as a major cause of disturbances until performance improves. In any case, do not bury the problems in inventory and schedule change systems. A production network, both internal and external, well short of part-per-million defects should never appear to run smoothly at all times. If it does, a facade has been erected to hide the fact that continuous improvement is no longer a goal.

As for timing the deliveries, in the job shop case, give suppliers a time window in which to make deliveries of materials that inherently cannot be used in any repeating pattern. Hold to it. Use suppliers who can be developed to hit this window with good-quality material.

In the uniform load, repetitive case, convey to suppliers the upcoming schedule segments indicating the daily quantities required. Put them on a suitable form of pull system if they are ready for it. A card system works if distance and transport delay are not great. Where the time for cards to circulate back to the supplier is a matter of days, an electronic signal is preferable. It cuts out the return delay time.

A unique stream of material that is to be sent in sequence can also be conveyed electronically. An example is car seats delivered to auto final assembly. A computer can signal to the seat manufacturer a few hours ahead the unique sequence of seats to deliver.

Standard Containers

Each standard container holds the same number of each part. If the containers are celled or egg-cartoned, part counts can take place as they are filled with minimum mental effort. Counting parts loaded in standard containers means only counting containers.

The same size standard container should do for as many different part numbers as possible. If standard containers come in few different sizes, then most are the same size and standard material handling methods and equipment are possible.

Small containers are better. Should a process go sour, they restrict the inspection to a small number. They can be manually moved if necessary.

Small and standard size containers better serve as "inventory clocks." A guideline for using them this way is to put no more than one-tenth-day's use in each container. That makes a more finely calibrated "inventory clock," and adjusting an upper limit on inventory of a part can be done in small increments.

Containers for external transport are subject to other considerations. What size fits common trucks and material handling equipment? Should they be returnable? (In many cases, yes. The properly labeled empty container is a signal to fill it up again.) Should an industry have a set of standard containers? (Not a bad idea. It promotes material handling and transport standardization at suppliers common to an industry.) Should the containers be covered or otherwise specially designed to protect contents? (Sometimes necessary, but covers are extra material handling, and special protection is nonstandardization.) Should containers be collapsible? (Prevents "backhauling air" in returnable containers, but they take time to set up and collapse.)

For smooth, repetitive production, standard containers are a given. The considerations on type have no absolute solution, but the issue is not trivial. One major expense of transport is containers, and that is even more true of inventory sitting. One can mark some plants who have "gone JIT" just by driving by and seeing stacks of no longer used containers sitting outside.

If a supplier and customer progress well, the last operation at the supplier should fill the customer's containers, which are then transported directly as possible to the customer's first operation.

Part counts are simplified when receiving standard containers. Count the containers, assuming they must be counted. (The assumption should be questioned.) Sometimes counting is simplified by bar coding. If the containers are arranged in a standard pattern on a pallet, the count is even more simplified, and anything simple and effective cuts waste. Trimming inventory to the bone may even eliminate the receiving count. The customer simply pays for the count of parts used, and any inventory received but not used is too trivial to worry about.

In addition, transport time and expense are saved by not only handling standard containers at standard times but by simplicity in tracking, tracing, and routing. Sorting irregularly packed, non-standard material in a truck terminal is expensive and subject to routing errors. Standard handling eliminates time and expense at supplier shipping docks and customer receiving areas. Standard handling by the same people also eliminates much shipping damage and prevents oversights. (A serious problem with cargo handlers unfamiliar with returnable containers is that they sidetrack the returning containers, not understanding that it is just as important to get an empty back as a full one there.)

Shipping

As in production operations, the more a company can capture repetitive potential in shipping operations, the better. If possible, establish "milk runs" between plants. A typical milk-run truck makes a daily circuit of three to five suppliers and brings a mixed load of material from each back to the customer plant. Or, the same truck might run a slightly different route each day, picking up from some suppliers daily, some every other day, and some once a week. The combination depends on weights, cubes, and distances.

This approach depends on distance because a truck with a few stops cannot make much more than 250 miles in one driver-day; thus if suppliers are spread out, the economics at best have to include an overnight stop in the planning (and a lone supplier in Glasgow, Montana, is not likely to be a feasible stop on any circuit).

If some suppliers are clustered at a distant point, an alternative is to milk run the pick-ups and combine their load into a long haul. It beats the delay of waiting for a full truck from each one. The Glasgow, Montana, case may not be hopeless either. If a trucker (or rail shipper) can be found who will combine a partial load with something else on a regular basis, even that situation can be improved without subsidizing a trucking company.

Distance is a concern, but traversing it is the business of traffic management. The real distance from a supplier is a question of the time required to transport material and the opportunity to work out details. Some electronic companies have flown circuits from the Far East on a daily basis. The principle of immediate feedback is preserved, and the transport cost as a percent of value is still low.

For heavy items carried by ocean freighter, transit time is weeks, and the problem grows. One serious proposal was to station a machining center for heavy castings on board ship so that value-added work was performed in transit. (It was rejected.)

Long transit distances generate two notable problems. First, protecting material for a long haul with multiple handling requires expensive containers, which cost. Second, time is a big factor: lead times to lock up schedules are extended, and flexibility is reduced.

For instance, Kawasaki U.S.A. brings in a mixed load of material sufficient to resupply each assembly line daily or every other day. It comes by sea from Japan—six weeks from packing to receipt. Schedules in Nebraska must be fixed about 10 weeks in advance to allow parts to be fabricated in Japan in a uniform load segment and shipped to the United States. Shipping time variance is not a big factor. Only one- to three-days' safety stock is maintained in Nebraska. Rarely has production been missed because of weather. Most managers in northern areas overestimate the risks of shipping delays due to weather and underestimate the problem of added lead time restricting flexibility and adding to overhead.

For example, several American electronics companies start material in the United States, ship it to the Far East (by boat), then to Mexico for subassembly, and finally back to the United States for final assembly. Total lead time: about four months. Long lead time materials planning with gross-to-net inventory calculations is inescapable. Much of the inventory must be tracked in transit, and part origin must be country coded for customs. Engineering changes are a logistics exercise, and responses to market changes are not rapid. Customs, inventory carrying costs, and transport expense are considered when off-shoring production, but the total effect on company overhead is not.

SCHEDULING SUPPLIERS: AVOIDING DATA AND ANALYSIS LAG

The more one understands about a supplier's operations, the better they can be scheduled. The prevailing custom is to accept the lead time given by each supplier and provide them with orders or delivery schedules giving them the best possible information within the lead time they ask. The longer their lead time is, the less valid the information originally given to them will be, and the more likely that it must be amended in some way during their lead time.

Check into the real times a supplier must make various commitments progressing toward delivery. A typical situation is trim piece sets ordered by a furniture manufacturer, a color-and-mix problem. Each trim set fits a particular model built by the customer, but in six different colors. The customer company only knows the exact models and colors it will assemble three days ahead, but the supplier gives a lead time of one month for trim sets.

The supplier needs to order material specific to each trim set one month in advance. However, it can fabricate the configurations of the trim sets two days before shipping, apply the final color one day before, and ship overnight. If the customer gives the supplier its final assembly schedule three days ahead, the supplier will deliver the correct trim sets on the morning of assembly (provided it had the correct material on hand).

Now the supplier can order the correct material a month in advance no more accurately than the customer's latest forecast information permits, but the real purpose of ordering material is to be in materials position for a range of target trim mixes a month later. The customer must at that point know the model and color mix target range it wants to be in a month later. Customer and supplier taking the same position makes sense. Then the furniture manufacturer can make up three-days-hence final assembly schedules with little fear of a "Can't do" phone call from the supplier, and without scanning supplier inventory data.

The supplier, when investigating its own suppliers, may in turn find a way to cut lead time of final commitment to raw material for trim parts. If the furniture manufacturer develops its schedule into uniform load segments, planning and execution are assisted for both itself and for everyone in its supply train, provided all parties learn how to work with it.

This example illustrates several important points:

1. Supplier and customer work operation to operation as nearly as possible. One supplier from one location delivers trim in sets to the customer location of use. Simple planning. Simple communication. Simple execution.

2. Commitments are made at the right time from points close to the action, and they are communicated nearly directly from action point to action point immediately. Every morning the supplier receives a three-days-hence schedule from the assembly process doing the work. The one-month materials position directive (forecast) comes from the plant. *Its system eliminates most planning cycle lag and communication time lag.*
3. No second-guessing of plans takes place. The more people involved and the more stages in planning, the more likely that is. Supplier and customer work as a unit.

Computers and communications technology do not work this way by themselves. Masses of data bring no profit except to the computer and communications companies, so pride in pumping a lot of data is misplaced; but computers used well are important tools linking the total manufacturing network.

American auto companies exemplify the problem of not communicating point to point, though their systems are probably better than most other manufacturers. In Detroit they centrally process schedules for suppliers to assembly plants. Weekly update "releases" are sent from there to all assembly suppliers, some by mail, but most electronically. To do this, Detroit must know the inventory position of each assembly plant and approximate its upcoming assembly schedules.

However, each assembly plant sets its own assembly schedules in final form three to four days in advance, conforming as much as possible to original plan but considering immediate factors. To any individual supplier, material actually used in a given day or two of schedule can deviate substantially from the last "release" from Detroit. (Many of the suppliers are internal, too.)

Recently, assembly plants have begun to send daily "JIT releases" to selected suppliers. These are basically demand replacement delivery requests based on their own final assembly schedules. A few suppliers may be placed on card-type pull systems—demand replacement methods. It can be done for nearby suppliers developed to work operation to operation.

Data and analysis lag plague many systems that are otherwise computerized marvels, much as in Detroit. By the time data is collected, goes through its processing and review cycle, and is transmitted to the point of action, it is already dated. More than one auto supplier has concluded that it would be better off if it just had access to the order banks (backlogs) and final assembly schedules of the plants it serves. Some partially circumvent the system with telephone contacts in the plants.

This is second-guessing a forecast when the guessers would like to know what is really happening with certainty. People near the action point working with trusted signals of actual demand are not much tempted to second-guess. Point-to-point communication of actual demand or demand replacement eliminates most second-guessing at this level.

The purpose of central planning is to eliminate the second-guessing of an overall plan that must be based substantially on forecast because of long lead time. Considering the pronouncement emanating from a central planning area as a *commitment* and not a forecast also stops some second-guessing. Handing the "commitment" to several tiers of suppliers from one central point at one time eliminates cumulative replanning delay (often a month for each) resulting when each tier does it independently. Also, each supplier tier adds more people to do second-guessing. All that lead time and planning churn is buried in inventory.

The principle: Plan centrally with dissemination simultaneously to as many layers deep in the supplier network as feasible. That is working from the decision source—the plan. Communicate the actual as immediately as possible from operation to operation, wherever they may be located.

Such a network is impossible for a horde of suppliers, many of whom two tiers or more down are not even known and who change frequently. It starts to become practical for a small network with long-term working arrangements. (A large multiplant manufacturer is such a network. The principle should apply there also.)

Nonsynchronized planning cycles among companies and suppliers create static in a network. Some do a major replan monthly; some roll a plan weekly. Monthly replans do not all process simultaneously. Some are complete by the 10th of the month before; others not until the 28th. Shop calendars do not all coincide either. No supplier can establish a replanning cycle that avoids use of stale data from some customers, which may not be undone by fresh data arriving soon after.

No network planning in the United States is as synchronized as the automotive supply networks of Japan. All major Japanese auto companies roll major plan changes monthly and minor ones weekly on the same calendar. All majors complete planning for the following month by the 20th to the 22nd of the previous month. First-tier suppliers using fresh data can finish their plan for the following month by the 25th, still leaving time before month-end for second- and third-tier planning. In addition, second- and third-tier companies can access the top-level plans as soon as available if that gives them a headstart in their own planning.

Reasons for nonsynchronous planning cycles may be happenstance. Until someone starts to build a supplier network, the problem does not enter consciousness. When it does, a first consideration is the total business of each supplier and its purpose in the network. Many are viable only because they have nonnetwork production. Suppliers' own planning systems must consider all customers, not just those with a close tie. Different industries served have their own problems with planning cycles and lead times.

This problem can be minimized by suppliers designing operations in the service of particular sets of customers, even if two different operations make almost the same thing. Planning cycles are seldom considered in thinking about why and how one focuses a factory, but they must be as companies try to link operations more closely together. The scheduling issues appear to be hokum unless people and physical processes are developed to be executed by them. Scheduling, people, and process develop together.

In the past 10 years, American companies have made admirable starts in this direction. Many give advance forecasts to regular suppliers. Many "buy capacity" from suppliers on a long-term basis, then firm the scheduling of that capacity as time arrives for detailed operations planning. Muscling up to much more than that is a supreme challenge in drawing together a network of organizations in business for themselves.

STARTING WITH SUPPLIERS

Early fascination with JIT often stems from supplier delivery, one of the outward signs of its existence. Then, managers press for precision delivery without developmental groundwork in quality, processes, schedules—and, most of all, people. Developing the external factory is an extension of developing the manufacturing proficiency of the company internally; but, molding independent business organizations into even a loose network is more difficult than internal reform—challenging as that may be. Hardly anything bucks the traditions of American business more. Every company seriously striving for manufacturing excellence has found that to be true.

Start with changing the purchasing manager's role from adversary of many suppliers to developmental coordinator and coach of those who can make the cut on the team. Some managers think this means going soft on suppliers, and occasionally it does. A coach must explain what the suppliers do not understand, but coaches are not always easy souls when stressing performance. As with any team, one would like to cut some players but cannot. Those are in need of coaching but cannot accept it.

First, develop those who must develop the suppliers. They cannot coach what they do not understand, and while some of that can come from a training room, some should come from internal experience.

Purchasing managers are not the only people needing training for external contact. Engineers, production controllers, traffic managers, quality instructors, and others who work with suppliers need preparation for it. Each supplier company is a different environment, so this is something like preparation for consulting.

Then, start wherever you need to start. That depends on the specifics of each company. Usually the starting point is quality. Then, as internal schedules develop consistency, suppliers can be held to tighter standards of delivery, progressively working a number of them into "demand replacement" delivery where repetitive potential is developed.

Early on try to form some kind of association with key suppliers. This is a valuable forum to obtain supplier participation in forward engineering of new products. From this association come big cost savings and competitive advantage in cutting the lead times for introduction of new products.

CHAPTER NINE

Taking It to Them—Marketing Power

All this would be beautiful were it not for the needs of real customers, or so it is feared by marketing representatives. By their instinct, if inventories are reduced, surely customer service will deteriorate. The promises are seductive but with little assurance how soon, and meantime a disaster might occur.

With the exception of champions of change, almost everyone fears this, so companies temper their adventuring with self-imposed marketing restrictions: Never shall customers be worse served than at present. Improvement targets shall include customer service—things a customer can see. Whatever pain the organization must endure for internal reform, the customer shall see none of it.

To allay marketing fears, companies safeguard customers from operating changes at first. Finished goods inventories temporarily increase until customer service from lower inventory levels can be demonstrated. Make-to-order product may actually be built ahead to demonstrate shorter lead times, then held for shipment. Extra inspectors check final product to be sure that responsibility at the source is working.

Sooner or later, marketing must decide that it is part of the movement to manufacturing excellence—and the sooner the better, because marketing is vital to it. It can no more afford to be independent of engineering and production than those two functions can be independent of it. JIT/TQ brings changes in how to think about marketing.

Sales representatives often think of themselves as selling product from a shelf, whether literally true or not, and the job of the operations side of the organization is to keep their wagon full. Then salespersons need little concern themselves with company operations to make a sale. Attract interest, overcome objections, and close. Here it is and there you are, and if something is wrong with it, we will exchange it.

Confidence in selling goods that need no exchanging from a wagon less full comes from confidence in the total manufacturing organization. Sales representatives match more than an inert product to customer need. They convey the total service of the organization, of which the product itself is a part. In the end, everyone is in the service business.

Marketing is more than selling. Marketing is establishing what the customer wants, but this is useless without combing through the detail of *how* to provide it. Marketing begins with a study of the customer, learning the customer's need better than they know it themselves. Success comes from working out in detail *how* to fill this need, and the how has much to do with determining strategy.

To develop a marketing strategy, one needs to understand how operations can reasonably be developed to support it. Thinking of the customer comes first because smooth operations are of little consequence without delivering what the customer wants, but a great void in operations ingenuity likewise results in mere dreaming about a competitive marketing plan. The marketing side of a manufacturer going into JIT/TQ should begin to evaluate the possibilities as soon as possible.

Few American companies entering JIT/TQ have begun thinking about the marketing aspects soon enough, but this appears to be a major key to long-term success. Consider a company making a product over which production control is so inconsistent that final product must be graded and sold by grade. Examples are electrical resistors, certain bearings, some kinds of polymers, and semiconductors. Suppose the first objective is to develop consistency in production so as to hit target grades more consistently. Yields improve, and eliminating some of the low-end grades soon seems feasible.

What is the marketing effect? First, the company should better be able to deliver target grades on time, particularly at the difficult end of the spectrum. Then it can think about upscaling the grades actually delivered. Offer a medium grade for the former price of low grades. Increase the percentage of high grade sold. Finally, perhaps it can think about restructuring the grades offered so as to obtain a strategic advantage. Simply eliminating some of the grades might be a very good marketing move, cutting down the choices necessary by customers and the haggling over them.

This kind of situation often seems to become stuck in marketing neutral. The marketing side of the company establishes grade ranges and prices them. Its sales projections are stated in terms of them. Production is busy giving marketing what it asked for, which apparently satisfies the customer, and everybody loses sight that a better world just might be possible.

A corregated box company began thinking about providing quality JIT service to a variety of customers. A few of them were beginning to ask for it. The company's operating people could foresee considerable improvement in quality, setup times, lot sizes, and deliveries, but it immediately became obvious that the traditional sales strategy would no longer be valid. It had specialized in large, low-price orders wangled by methods occasionally employing football tickets and Christmas booze.

The sales force could no longer be box sales people but rather service representatives in the true sense—analysts of each customer's operations. How did the customers schedule? At what rate did they use boxes? What did quality mean to them? Should the company target small, short lead time orders as well as large ones?

A few of the box company's biggest accounts were handled personally by the general manager. That practice would probably continue, and a few accounts might even be added, but what role would the general manager play? What would be the function of a service representative in a long-term, operation-to-operation agreement with a customer? Though not completely defined, these roles cannot be played competently on the basis of vague familiarity with either customer operations or box plant operations.

Allen-Bradley has developed a well-reported example of computer-integrated manufacturing using many JIT/TQ concepts. Contactors and relays for industrial motors are assembled on an automated line, where much of the equipment is Allen-Bradley's own design. A large variety of models and variations are assembled in lot sizes of one in random order, which is accomplished by bar coding jobs and parts. Sensors reading the codes trigger the equipment to download the right programs and select the correct parts. The line must run a little below maximum speed to balance the random sequence.

The marketing side of this story is interesting. Until April 1985, Allen-Bradley made only American-style contactors and relays at the Milwaukee plant, but the devices used for European and Japanese motors are smaller and conform to International Electrotechnical Commission (IEC) standards. The world market, including the American portion, is moving in this direction. The company set its sights on more than just hanging on to U.S. production of a declining segment of the U.S. market. It went for world market.

Allen-Bradley decided that assembly in lot sizes of one with 24-hour turnaround was necessary to be competitive. Same-day production is an erratic load but, fortunately, only a fraction of customers ask for next-day service, so the backlog can be used to somewhat level production. Allen-Bradley is capable of receiving a foreign order by wire in the morning and, if it can catch the right airplane in the evening, delivering to Europe or the Far East within 24 hours, or close to that.

For the moment, Allen-Bradley is the world's low-cost producer of these items, plus the factory service is equivalent to off-the-shelf delivery. Two marketing punches beats one.

These examples have all been of product lines primarily supplied to other industries. Each company has potential for being meshed into the production networks of several different customers, thus its marketing outlook would take a different turn with that, but each also has marketing potential not going in that direction.

The marketing potential of excellent production has been amply demonstrated by the marketers of imports, notably Japanese, but also many others. The attraction of imports is either cost or quality or both. They can seldom compete on delivery (except in an Allen-Bradley case), but keep in mind that such a thing could be done in reverse from any country having good air connections.

The market potential of well-manufactured imports in service and consumer markets is likewise well known. Price alone will capture buyers once, but if a product also possesses quality, unique features, or complimentary service advantageous to a market segment, that will be decisive. Go for as broad a spectrum of advantage as possible.

Capturing several advantages at once is the dream of marketing strategists, but actually pulling it off requires operating capability. Strategic plans without ability to execute are specious. Marketing plans must include the operations insight to formulate something simple and realistic and the leadership to bring it about.

As long as breaking the traditions of manufacturing is seen only as the province of production and engineering (who, hopefully, will not offend the customer when doing it), the movement is only modestly supported—beset by doubts and fears. As soon as the activities moving toward manufacturing excellence are coupled with a strategy to capture and hold market, the company is united behind a driving force.

Thus far in the United States, marketing departments have been slow to see this, and so JIT/TQ (or whatever the company calls its movement) remains mostly a bubble-up-from-the-bottom situation. With JIT/TQ, as with everything else in business, the customer comes first, and nothing becomes powerful unless it is connected to a market incentive for performance.

THE MARKETING HOLE

One of the most serious early problems among American companies most successful operationally with JIT/TQ has been marketing. If a company improves productivity 20 to 30 to 40 percent but does not increase the volume of product to manufacture, attrition cannot offset the excess people. Omark Industries has suffered the problem. Harley-Davidson has. Black & Decker has. And all of them started with the intent of treating their people as valuable resources not to be lightly dismissed.

The conventional solution to overcapacity is scaling back: dismissing people, selling equipment, and closing plants. After a company has made a commitment to help its people and to regard them as its most valuable resource, such a decision is made with great anguish; but if no alternative appears, eventually it must be done. Productivity improvement does not automatically reward a company with increased market share to take up the slack. Another way to look at it is that marketing strategies were not in place to make use of the resources and therefore, as part of its commitment to manufacturing excellence, a company must very early in its progress become aggressive in seeking market. It must dig out of a marketing hole.

This is a formidable marketing challenge. Products compatible with a company's production capability are not necessarily compatible with its marketing capability. Rummaging for something to make is no way to go about it.

Contract production seems a quick solution. The customer does the marketing. However, excellent manufacturing requires well-forged linkages among marketing, engineering, and production, and thus the problem becomes finding an understanding marketeer. Many prospects believe that contract production will somehow keep their wagon loaded with no special participation on their part. Finding an understanding partner is a chore.

Do-it-yourself marketing takes two routes: (1) increase market share for current product lines and (2) introduce new products. Both routes have hazards.

Increasing market share is easier after quality and cost improvement markedly enhance an old product line's competitive position, but delay factors are at work. Substantial results may lag the onset of productivity improvement by many months. Meantime, the marketing expense to hold share of a weak product line saps resources that might better be conserved to build awareness of product advantages when they arrive—if one has faith and cash flow long enough. No prescriptive answer automatically solves this classic dilemma.

A new market for a new product is a venture into the unknown. Edsels and Videodiscs sometimes happen despite pre-testing. Should the new product not fit an established marketing channel, developing it may unsettle existing channel connections. Deep-pocket companies usually avoid markets they little understand, selling off product lines with the terse explanation, "Does not fit the business." Shallow-pocket companies can find themselves betting the store.

The safest approach to new product development follows a sequence of thinking that begins with customer need then evaluates whether it can be served through established marketing channels: Need. Channel. Design. Produce. Thinking of production capability first could hardly be a more backward way to seek something that will fill a customer need. A company can find itself trying to bring back the cast iron bathtub.

The company lacking verve for market gambles will ride a tired but now well-manufactured product line into a slower oblivion. Some of the money taken from inventory should go into product R&D and marketing.

The opposite tack is to go for a product so hot that the major production concern is producing it fast enough. Forget the manufacturing excellence tedium. High-tech-and-promote the return to stardom. The danger is in setting up the company for return-mail attacks by excellent manufacturing competitors capable of very fast downward learning curves. Company position could fade fast as products edge toward maturity, the stage in which refinement and continuous improvement capture a big advantage.

Solid manufacturing survival into product maturity depends on coupling well-designed total marketing approaches with manufacturing excellence—disciplined, full-scope innovation. Allen-Bradley's connector and relay business has elements of this approach. American Hospital Supply has an element of it. This company offers not only "to sell" items to hospitals but to monitor their inventory and resupply at the right time, a tactic considered controversial and possibly illegal in the eyes of competitors, but it probably presages things to come.

These are company strategies, not marketing strategies or production strategies. To formulate them, managers should consider "supply side" marketing strategies derived from insight into manufacturing possibilities. The strategy is made stronger by combining these into a simple, unified plan of attack.

INNOVATION

Product life cycles interrupt the continuous improvement process in manufacturing before it reaches its zero-everything ultimate. Some changes are leading-edge technology, others but fashion changes. New technology obviously affects the production process. Marketing shifts do also. A decrease in promised lead times, a change in marketing channels, a push to heighten interest in options—all such changes in total market package affect production.

Innovation is successful change—much more than invention, which is demonstration of workable hardware. Innovation is acceptance by a market segment through creation of a total market package offering and development of an organization to serve that market.

Few inventive companies (or people) are considered failures because they could not produce something—somehow. Failures are ascribed to something else: Basic concept flawed. Cannot sell. Cannot manage money. People trouble. Not enough capital to survive the inevitable calamity.

These postmortem analyses designate several single-perspective weaknesses, but in any case, the decedents were "unable to put the whole show together." In world-class manufacturing, a major backstage success factor is ability to keep a flexible linkage for change between marketing, engineering, purchasing, and production.

Every manufacturing company has such linkages, and a primary one is the engineering change system or its equivalent. However, after product lines become old and full of history, the linkages become arthritic. Changes lock up in their own administration. For JIT/TQ to be effective, they must be loosened, and, as discussed earlier, an effective way to do it is by putting the change cycles on a regular pattern (such as once a month) and clearing away the inventory and lead time problems that obstruct changes.

The implications for product improvement are great. The same regular pattern of change should be equally effective from that point of view, but this depends upon having a regular pattern of feedback information from field sources and a short-cycle review time to feed it into the change system. The regular feedback from field sources is the tough part. Many feedback mechanisms are designed only to report complaints. Those are important but not more so than ideas for improvement of the product in use. Garnering these and assembling them on a regular basis is not a novel idea but the original thought behind model-year marketing.

Major innovation is a discontinuity in past improvement trends. Not only is the technology subject to radical change, but so is the market. A perfect example is IBM's entry into personal computers. For IBM, the technical change was less radical than the change from building mainframes to order to building PCs to stock—some technical change, but a big customer change and a new world of channel building and logistics to work through.

They spun off a "skunk works" project to develop it. To break out of the past, a new, small organization had to start back where it was simple again and recombine the company's talents in a new form. Linkages between the functions had to be very tight, and that was easily done by keeping the organization small and free to redirect how everything should work.

A new word for this is *intrapreneurship;* but, like most management thought, not much is new in the idea. The advantages of a small, innovative enterprise sound much like the precepts of continuous improvement: high internal visibility. Fewer people, fewer molds to break. People wearing multiple hats have a broader perspective. If the people in the "skunk works" group have the concepts of JIT/TQ in mind from the beginning, they should form a powerful organization for a new venture if their strategy for it has no fatal flaw. "Skunk works" spin-offs sometimes fail for a variety of reasons but, in one form or another, they are useful when making a break with the past.

A first impression of continuous improvement manufacturing is that it will stifle market innovation. Not so. A skilled manufacturing organization should acquire new technology, engineer more practically, and start a new product further down the learning curve. It should then progress rapidly from there.

Naturally, there is a hitch. One must learn how to do it, which is both an intellectual and emotional experience. Process development anticipates new products, and new products are developed with produceability in mind. Continuous improvement is disciplined innovation in the sense of going for a new product concept with as many market advantages as can be mustered. That game is not mastered by a collection of brilliant individualists pulling in their own directions.

Along with proficiency in introducing new product lines comes good judgment in how to handle old ones. Old products built with old tooling and equipment require old tooling and equipment to be kept and maintained. Pruning the product line at judicious times helps keep the newer products competitive. If occasionally providing an old design is vital to customer appeal, this capability and its cost must be carefully evaluated. Perhaps an "off-line" production facility for relics can be set up or the problem otherwise solved in some such fashion.

MARKETING'S ROLE

Besides expanding the scope of marketing strategy, manufacturing excellence is apt to require two major changes in performance of the marketing function: (1) constant feedback of customer problems in such a way that engineering and production can overcome them (part of total quality), and (2) managing demand so as to sell from production capability rather than from stocks and backlogs, which buffer the customer from production. Marketing representatives are typically unaware of this when production begins its reform. They need to be a part of the revolution from the beginning.

Staying Close to the Customer—First Step in Quality

Suggesting to sales representatives that they are not close to the customer will bring well-deserved howls of protest. Suggesting that engineering and production are not close to the customer is apt to get their instant agreement. Suggesting that it is marketing's job to bring the customer close to engineering and production is likely to evoke the response that sales representatives certainly try, but the clods do not wish to listen.

Defining customer need is basic to marketing. Everyone does it—knows that they must—and yet the means by which customer needs are translated to engineering and production is never as good as a company would like. Most manufacturers do not have regularly and systematically organized information to guide the company in two areas where it really counts—engineering and production. Without it, although the sales effort may be close to the customer, the *company* is not.

In such industries as packaged consumer goods, market research has matured so that everything from package design to production rate is at least partly based on statistically summarized evidence of consumer appeal. In other industries, the skilled manufacturers have learned to incorporate market research into the development of new products. Designs are refined using simulations and tests guided by experiments statistically designed to extract the most information for the least cost.

The weak area for many manufacturers is thorough follow-up of a product in the field to correct the little irritations to users of the product. The difficulty of trying to observe actual customer use and the sluggishness of the engineering change system in most companies bog down these follow-up efforts.

In addition, many manufacturers do not make maximum use of returned goods for intelligence. Engine or electronics companies whose reputation for reliability is critical do make extensive use of teardown analysis, but those with more mundane products may not. Reasons for return may be categorized, but perhaps not in a way to provide a regular stream of ideas for product improvement. A problem is referred to engineering only if it is "really serious."

However, a dictum of total quality is constant improvement, and this applies as much to the product as to the production process.

As is true of improving production processes, companies will improve products only in response to serious complaints—unless there is a policy of always looking for ways to make a product better. One reason for reduced customer feedback to engineering and marketing may be that these functions have historically been slow to respond to it. Fast action took place only if a corporate authority figure pronounced that either a product liability problem or a market disaster loomed unless a change were made. However, reducing inventory, simplifying designs, decreasing setup times, and all the rest should substantially lubricate the creaky old engineering change system.

A first impression of JIT is that for smooth flow, a product should run without change for months on end. Not so. Many engineering changes should and can take place for both product and process improvement. Just organize the changes so that they do not destroy production flow. Group the disruptive ones for execution at schedule change times.

Total quality begins with the customer. It is no accident that senior managers experienced with total quality stress this point. In fact, after production and engineering have progressed for a time with total quality in their functions, it becomes clear that perfecting the production quality of a less-than-perfect product limits progress.

Marketing does not just get a windfall from total quality by suddenly obtaining products that have better customer acceptance. They have a responsibility to assist the improvement process.

If engineering and production become more responsive to quality improvements in the products themselves, a constant flow of credible information about the details of product use becomes necessary. This information is not obtained without searching for it.

Many manufacturers have little direct contact with the actual users of their products. Some are suppliers to other manufacturers: the makers of disc drives, for example, may supply computer manufacturers who in turn sell the drives to purchasers who still may not be the end user. Other manufacturers sell through independently owned and controlled distribution channels. Unless measures are taken to obtain some feedback first hand, all user reaction to the product is filtered through all those intermediaries.

These filters in the feedback chain should be a concern not only to the manufacturers, but to the marketers of their products. A manufacturer who does sell direct to users should think about this in establishing close relations with key suppliers. Part of improving total quality with them is forming a method for participating in whatever feedback can be obtained from users. The idea of actively *developing* better quality is a very different outlook from that which merely searches for something hoped to be better from a different supplier.

The necessary shift in attitudes to do this between a manufacturer and its suppliers is not so different from that which is necessary between marketing, engineering, and production inside a company. The task is not trivial because it requires organized cooperation.

Feedback from users is neither a new issue nor an easy one in marketing. Special personal contact with users is expensive, and may even rile them if contact is made at the wrong time. "How-are-we-doing" cards bring well below a 100 percent response rate, and interpreting them is not always straightforward. Feedback from employee users of a product also has its biases, for example, knowledge of product construction.

Total quality increases the heat on this issue. In the end, for a product whose use should be better understood by the manufacturer than the user, there is no substitute for knowledgeable people directly observing users. A knowledgeable person understands the product technically and perhaps is even familiar with the process for manufacturing it.

Some of the difficulties in obtaining good feedback on customer problems involve the quality of technical communication; many more stem from reluctance to make attitude adjustments. Sales representatives may not be very motivated to spend field time working toward improvement of products rather than selling them. Pay that is highly based on commission does not encourage it. Likewise engineers remote from the customer may not accept that *their* handiwork is the cause of a user problem until they see it for themselves.

The implications of this for training sales and service representatives is considerable. All that was mentioned in earlier chapters about developing production operators to be good observers, use systematic methods to troubleshoot problems, and so on—all that applies to sales and service representatives also. In short, *everyone* in a company should receive regular training on total quality.

Sales and service representatives regularly in contact with customers (and users) are in the best position to systematically gather information about product problems. However, much good comes from having engineers do this on occasion, and many companies make this a practice. Sometimes great benefit comes from even having a production person discuss product strengths and weaknesses with the users of a product. Breaking down the walls between departments is a prerequisite to improved *company* contact with the customer. It not only improves the product but also the service.

One major source of customer complaint is the service offered with the product rather than the product itself. Customers are by no means perfect, and some are unreasonable, but service quality should be pleasing to most of them. Quality demonstrated in the customer's presence is more visible.

Quality of service is a big factor in marketing, and quality is important in any form of customer contact, including that intended to elicit information on how to make the product offering better. Grass-roots work in understanding the customer is poorly done without careful preparation for it; but, by avoiding it, chances are greater that expending great effort elsewhere in the company will not be appreciated where it really counts. Total quality begins by understanding the customer need for the product in detail.

By coupling the methodology of total quality with the need to visualize new markets among a group of potential users foreign to the company, one can see that a formidable challenge awaits marketing.

Managing the Flow of Demand

Prompt knowledge of actual demand is invaluable to any marketer. This is the marketing benefit of the practices of demand management outlined in synchronizing the schedule in Chapter 8. A long marketing channel inhibits many manufacturers from knowing if a sale to the first stage in a channel is resulting in a sale to the final user. Low inventories and ordering by demand replacement helps to assure transmittal of actual demand to the manufacturer.

Marshalling demand into the time segments useful for production scheduling is possible in a somewhat mechanical way in the make-to-stock case. Periodic reviews of finished stock and ordering time-coverage quantities do much of it. In the make-to-order case, managing a backlog is less mechanical. Industry custom and competitive practices do not allow total firming of orders.

Suppose a backlog consists of items normally sold on bid, so that outstanding quotations are a preliminary backlog, but no one can be sure which orders will actually be built until quotations are accepted, and then there may be cancelation clauses. Suppose a bid acceptance arrives late. Is the delivery date delayed? Suppose competitors regularly lie about delivery capabilities, but the lies often accomplish their purpose by getting orders. What do you do?

Three situations recur over and over when managing backlogs. Orders assumed firm may be changed in quantity, specification, or delivery—or canceled. A backlog is less an orderly database than a collection of live customer interactions, some of which squirm and wiggle. Aligning them into nice orderly time segments for production requires resolution in marketing policy.

The issues revolve around doing the right thing by all customers versus straining to accommodate some indecisive ones and thereby impairing quality service to all. The basic issue is common to all businesses: To what kind of customer do you cater?

Some years back, a restaurant down the street went through three owners in a year because each had little resolution to stick with their marketing policy. The restaurant was ideally situated as a family restaurant, but all three owners served liquor, which might have worked out had any of them turned away a set of people who visited occasionally. This group regularly drank more than they ate, becoming loud and obnoxious in the process. The patronage from true family dining quickly evaporated for each owner. (The successful owner's policy was to limit the liquor to two drinks and to every day attract one new customer who would return. In a year, that is a lot of regular business.)

Issues are a little more complex in a manufacturing company, but marketing resolve is equally important. What is an unreasonable cancelation? What is an unreasonable request for changing a specification? What is the policy on quoting and holding lead times? If the company can maintain short lead times internally, it can give customers short lead times, but only if the firm backlog is stable enough to actually make good on its commitments.

If a company is to deliver what the customer wants on time, it must have policies for establishing order parameters into firm commitments on the part of the customer. A good marketer can explain that quality service goes to quality customers. Marketing's role is to stay currently appraised of who the quality customers are, for these things change from time to time. One factor to check is the quality of the customer's own operations, whatever they might be.

Finally, one of the unsung factors in attaining manufacturing excellence is good customers. A company will not serve any customer well if it cannot organize its operations for it. In the process of doing so, give priority to the customers who will appreciate the effects of excellent operations. Let the competition go broke trying to serve the unreasonable few.

CHAPTER TEN

Reforming Permanently

Most beginnings are tentative: a layout change, a setup reduction, a statistical measurement of a process—a few modest steps. Most executives want to see something work on a small scale before laying company and career on the line for a cause vaguely understood.

From this beginning, middle managers are able to create pockets of improvement, if not excellence. Improvements based on technique run their course until new management ideas collide with old. Company reform becomes necessary for further progress, and that is a much greater challenge.

This area is not well mapped. A few American examples and the Japanese experience can be studied, but few Americans have the patience to exactly take the Japanese approach. The goals and physical changes seem simple enough, but integrative thinking and responsibility at the point of action are not small changes in management thought. Much business practice is based on independent competition by individuals, each in business for him or herself, so that companies are more coalitions than teams. The people side seems hopelessly Pollyannaish to anyone who has long lived in dog-eat-dog business environments.

Reform starts by a vision for it—an eye for true value added to customers and for cutting the waste in providing it. Think in basic terms. What provides value added to the customer, in detail? What is the simplest way to develop a waste-free total process to deliver it? Thinking this way begins to cut through all the side issues of automation, systems, several sets of accounting books, comfortable practices—all the encumbrances that distract a manufacturing organization from concentrating on specific missions to specific markets.

The reform is of operating companies—strategic business units that may or may not be independent financial entities. A conglomerate manufacturer must recognize the separate mission of each operating unit and regard each as a separate set of physical arrangements in the service of a specific market mission. Uncomplicate things.

Mentally, the most difficult part is integrated thinking. Ideas from here and there may be useful if they fit together, but great effort must be made to keep everything cohesive. Begin with a mission to the customer and piece together techniques appropriate to the mission. Done well, everything should link as tightly as pieces of an interlocking jigsaw puzzle. Some of the principles of thinking are summarized in Table 10–1, and they are not bound up in specific techniques. No such list can be complete, but it does indicate the necessary management mindset.

Emotionally, the most difficult task is motiving people to reform. Reform is confusing. It displaces people from comfort. They must be honestly motivated to reform. Harangues on a survival theme soon sound like a Chicken-Little routine unless the situation is genuine and perfectly clear. Motivating through strategic purpose is much more positive. Americans long to start something new—or what appears to be new—and "take it to them." Motivation comes from a shared vision of what *an operating organization must accomplish.*

In any case, managers and owners motivated *only* by money have little stomach for this. They prefer to sell trouble, to buy success, and to seek victory without training for it. If all constituencies are motivated only by money, the bond between them is weak. (Constituencies: owners, managers, employees, suppliers, governments, communities.) Each wants a share of value added as if they were splitting a jackpot, and they have trouble forming a tightly coordinated operating organization. Leadership by strategic purpose is important in establishing a working bond between all constituencies.

Manufacturing excellence is not an easy-money game. It is work. It demands improved performance by all constituencies of the company. Corporate leaders must be aware of the effect of their actions on each constituency because all have a part. Even a chief

TABLE 10–1 "Principles" of Manufacturing Excellence*

1. Take a broad but physical view of operations. Look for the possibilities, but techniques and technology exist to serve goals.
2. Maintain an active program to understand customer needs and desires in detail. Base improvements on that which adds value to the customer.
3. Eliminate any activity that does not add value to the customer. Seek and destroy waste in any form, great or small.
4. Make problems and conditions visible to everyone. With visibility, foster an atmosphere in which problems are openly seen and addressed.
5. Seek simple solutions. Keep it integrated.
6. Reduce variance in processes as much as possible. Measure it, track it, and set goals to improve it. Drive defects to zero or as close to it as possible.
7. Stop processes to stop defects if necessary, but try to failsafe processes so this is unnecessary.
8. Make maximum use of workplace organization. Use it to continuously seek ways to improve operations on the shop floor and away from the shop floor.
9. Keep maximum responsibility at the point of action.
10. Study and improve operations first. Then buy tooling and equipment as needed. (Avoiding automating waste, either with equipment or with systems.)
11. Create flexibility through employee cross-training, equipment development, setup times reduced for a lot size of one, and so forth.
12. Make maximum use of repetitive potential in operations.
13. Try to make only what is wanted when wanted.
14. Physically organize as many operations as possible for short lead times. Minimize dependency on long lead time forecasts.
15. Base as few decisions as possible on forecasts. Then, keep operations unified by working to the same overall position taken to prepare for a forecasted condition.
16. Once developed, standardize practices in order to hold gains.

*No list such as this can be complete.

executive officer must test the wind and sense the mood before setting direction. The leadership role in the beginning is more akin to leading a training expedition than a cavalry charge.

Total organization reform cannot be delegated. It is inseparable from the responsibilities of upper management. Converting a company that is lax, but does not realize it, into a taut, world-class outfit is challenge enough for anyone.

Generating the culture of continuous improvement among all employees and the other constituencies of the company is top management's responsibility. Continuous improvement is not a national or regional culture. It is a management culture, deliberately and painstakingly created by management. If upper management nurtures and protects the culture, technicians can do many marvelous things, but technicians cannot grow manufacturing excellence from a culture that rejects it.

GAINING DIRECTION

A well-crafted statement of business intent provides long-term direction. In an earlier era, Du Pont had a famous one: "Better things for better living through chemistry" (now shortened to omit "through chemistry"). The statement says what Du Pont does and implies what it will not do. Toyota's statement, "Cars to Love, The World Over," proclaimed its intention long before the company was an exporting giant.

Such statements are useful to upper management and boards of directors. They are also very useful rallying calls for all the constituents of an operating company. A company seeking to become a top manufacturer must have an overall goal that is easily understood by everyone—a reason for coming in early and working on Saturdays. Just saying, for instance, that a company wishes to become the world's foremost manufacturer of wood stoves is the first step getting there.

Plan strategy as if business were an extended war with numerous campaigns. What markets does one now have? What ones can be invaded? Which ones held? How? What competitive advantages in skill (marketing, engineering, production, field service, distribution) do we have now? What ones can we develop? Intention is constantly present, but tactical plans cannot be formulated more than one or two campaigns in advance.

A statement of strategic purpose is the objective of the war. It may be augmented by various slogans. A business plan would be based on this as sort of a war doctrine—an order of battle—advantages to be developed and weaknesses to be defended. For long-term thinking, an excellent exercise for a company having a strong foreign competitor is to assess what it might have to do to capture the competitor's market in its home base, considering basic operational capabilities as well as trade restrictions.

A company under siege must think defensively in its first battles. Pick them where defense is possible. After improving product and paring waste, is there still a chance? Situations that appear hopeless may, upon close examination, have potential if manufacturing can become excellent, but confusing bravery with stupidity is pointless. High-labor-content products with nothing else going for them are tough to defend in the United States. If at all possible, pick battlegrounds that lead to victory through multiple competitive advantages. Offense becomes a good defense.

Evaluate the competition carefully. A few American managers have said that they do not know how to evaluate foreign competitors not motivated in understandable ways. Ownership is not conventional. Some are owned by government agencies; some are owned outright by governments or combinations of them, so perhaps their motivations are merely to provide employment. Others are owned by interlocked combinations of banks and insurance companies. Financial results and the satisfaction from them are not easily comparable.

Not evaluating an enemy because it is differently motivated is a grave mistake. Strategic analyses based only on investment thinking sometimes assume that businesses will go through periods of investment, development, and finally harvest, coasting while doing so. They will be interested in keeping "good competitors" and will be satisfied with only a piece of a market that gives them employment. Suppose they are not so motivated but regard economic competition more as market war and keep on coming?

The United States is a global market for many products; therefore global competitors must be evaluated, and strategy must be to compete with the best in such an arena. Just marketing well is not enough. A top manufacturer must have close linkage between marketing, engineering, and production. Just engineering well may not be enough either. In the long run, a company must go for several advantages across the board.

One company's version of JIT/TQ is intended to give it market dominance in an arena in which it is already strong. For years, it has promised delivery of standard items (household products) off the shelf in five days anywhere in the United States. Now the company wants to ship custom-configured items anywhere in the United States in the same five days. It takes one day to receive and process the order and at least one more to ship it, so the plant has, at most, three days to line up orders, fabricate, assemble, and ship. Order processing and customer contact must be as failsafe as the work in the plant. The operation is developing into a manufacturing version of a United Parcel Service terminal.

This company is going for multiple competitive advantages: quality, variety, and delivery. It would like to not be far out of line on price also. (It still ships from stock, so it level loads the plant with items being made for stock.)

The contrast between two tough competitors, Honda and Toyota, is interesting. Toyota squeezes waste from every crevice—from design concept to delivery of product. By contrast, Honda squeezes waste down only to levels defensible against the Toyotas but strives for parity in total quality and superior consumer-engineering. Its total manufacturing process, including the supplier network, is geared to engineer quality products quickly. Both companies market to their competitive advantages while maintaining defensive strength in their weaker areas.

Creating strategic purpose helps a company attain the resolve to develop manufacturing excellence. Its thinking is clarified and the effort galvanized, but more is necessary. The grand strategy must translate into specific operational capabilities to support it.

This is the integrative part. How does every operational practice play a part in attaining the overall objective? For an inkling of this, go back to those dull, insipid diagrams in Figures 3–1 and 3–2. Until one begins to have an interest in how the pieces of the puzzle go together, they are uninteresting. With that in mind, they begin to show how a manufacturing enterprise is fitted together, bit by bit, to accomplish its purpose. In those diagrams, integration was from the quality perspective, but the idea applies to other perspectives as well.

Integrative thinking is progressing well when an individual production operator's locus of activity can be related to the whole and when the same can be done for the responsibilities of a sales representative. Objectives must be transformed into activities meaningful to the infantry soldiers.

REFORMING

Almost everyone will agree in general with motherhood statements on waste but have a hard time accepting that they are part of it. Opening the eyes to waste is a step toward acceptance, but persuading to reform is the first hurdle.

Personal futures seem in doubt so, from the earliest point of organizational reform, leadership and willingness to share downside misery become important. Reforming an old organization is tribulation unlimited. Evolutions seem slow, and revolutions are filled with bloodshed. Rather than go through this, management may opt to establish a new organization.

This is easier at the plant level than at the corporate level. If money for new plant or product is available, opportunity exists to start anew with different concepts rather than totally reform the old. The auto companies have tended this way. Hewlett-Packard attacks hardest on new product lines because its product life cycles are short. A production organization (and perhaps a marketing organization) can be reconstituted that way, but a company headquarters has little opportunity to sprout a "new philosophy version of itself" elsewhere. A company whose access to capital was limited by its financial arrangements, such as Harley-Davidson, had no alternative but to reform basic operations in place.

Rounding up a new organization and starting from a clean, simple position is a good way to minimize the pain of internally reforming an old, complex organization. However, breaking with the past through a new organizational and legal entity can be very difficult, as General Motors is finding with Saturn Corporation.

Reforming the old organization from the inside risks the whole thing occasionally erupting into the obligatory fistfight in the saloon (figuratively, if not literally). Changing people requires first-class leadership.

What about unions, especially militant, suspicious ones? Once a condition of trust and responsibility exists between managers and workers, unionization seems a minor issue, but living down a soap opera history of labor relations is no simple, handshake agreement. Sometimes it cannot be done.

Where this situation exists, many years' neglect of people relationships cannot be repaired in a hurry. Expect this part of the job to consume several years, during which shop-level improvements are modest. For example, when Mitsubishi Motors took over Chrysler's Australian facility, it found a labor environment as tough as any on earth. It took over three years, from 1976 to 1980, to prepare the human groundwork with the managers and with the work force. Much of the groundwork was daily education through workplace organization. Today, the same plant has attained quality and productivity levels rivaling any Western-managed facility.

The major problems are people, people, and people. Union objections are usually open and therefore recognized and sometimes covered by the press. Staff objections are quiet and subtle but equally difficult. No pat answer exists for dealing with people problems. They just consume time, patience, and energy.

Trimming waste and rigorously developing people are blatantly in conflict with assuring security to a major part of the corporate constituency: owners, managers, workers, and suppliers. Until crisis is unmistakable, they prefer not to budge from a comfortable status quo (and some not even then). An army distracted by splitting booty is difficult to keep in shape. Dealing with people who are challenged either by survival or by strategic purpose "to prove something to the world" is much easier.

The implications of this go to the heart of corporate thinking, a major perturbation of the usual economic assumptions. Taking a big profit not reinvested in the company distracts from rigorous conditioning because investors are not the only parties believing they should see some of the money. Maintaining any army's toughness is difficult in Fat City.

The leadership should have a multidimensional view of manufacturing excellence and what it takes to attain it. A single dimension of improvement (such as automation) is not enough and may not hold long in a level worldwide arena. Details of the vision may change from time to time, but as long as the overall goal remains consistent, overall direction will also.

Companies starting JIT and total quality wonder whether to make the internal change a separate project or just fold it into the regular responsibility of management. Project teams are very useful for definitive tasks, such as developing a plant to a milestone point where all work-in-process inventory is removed from a stockroom. A crash program to do that minimizes the time a plant must operate by confusing dual systems.

However, overall reform is in the hands of top management champions. They may form project teams for various defined purposes, but they must set overall tone and direction. That is easier if the reform goes through phases—a sequence of internal reform goals appropriate to the company situation and its strategic purpose.

Focusing on Primary Internal Reform Goals

Everything cannot be done at once. The goals of change depend on the current company situation, but, for most companies, emphasizing quality first is most appropriate. Primary concentration on quality may last a while, even years, because people and process development is not quickly done despite remarkable improvements now and then. The long-term, patient nature of the reform is probably best conveyed to most people as *eliminating waste to improve quality.*

A "normal" progression in goal emphasis might be something like this:

1. *Quality.* (Total quality)
2. *Lead times.* (Just-in-time manufacturing in the form applicable.) Strive for consistency first, study setups, and employ immediate feedback and visibility to assist quality improvement. Then, hold quality and strive for waste elimination by refining lead time reduction.
3. *Cost reduction.*
4. *Flexibility.* Some is achieved by reducing lead times and improving quality, but do not compromise quality and consistency of lead time by concentrating on flexibility exclusively. Add product complexity last.

There are exceptions to this priority list (first presented in Chapter 6). The important concept is unity of purpose through a progression of major objectives. Allen-Bradley, for instance, felt that it had to shoot for assembly lot sizes of one very quickly to capture a competitive advantage and stay in a market. However, it will be first to say that quality improvement is still far from perfect. The company has very effective quality at assembly but needs better than that.

For most companies, starting with a primary quality objective is best for several reasons. First, improving quality is hard to argue with. Few people are against it. Second, to make great improvement in quality, the company has to start with a review of the detailed need of the customer, which is a good place to start. Third, almost everyone understands that building quality capability is building long-term skill and expertise. It is not a quick-hitter program. Avoiding that impression early in the organizational reform saves energy for better things. Fourth, almost anything else the company would like to attain in manufacturing depends to some extent on mastering the ability to do it right the first time—quality.

Perhaps the most important reason for starting with quality is that quality improvement is consistent with a long period of human development. The whole idea of reform is best considered as a giant program of people development. Nothing can go faster than people (managers, supervisors, and workers) can work themselves into it.

For this to work, managers need to understand how various activities that reduce lead time and increase flexibility work in harmony with quality improvement—the start of integrative thinking. Otherwise, an operation with screw machines will assume that setup time reduction is hopeless and perhaps useless. (One idea is interchangeable sets of preset cams. Ideas *do* start once purpose and determination are in place.)

While developing the goals, keep thinking of strategy. What is necessary to capture or hold market? What do we do well now? What could we learn to do quickly? What can only come after long adaptation?

The reason for using any progression of primary goals is threefold. First, it helps to keep the effort from going in too many fragmented directions at once. Second, selection of goals should have as one purpose the progressive stimulation of a total manufacturing work force into higher and higher levels of performance. Third, organizationwide goals help to keep progress from being confined to the factory.

Other Goals of Excellence

Other issues of great concern to many companies have not been directly addressed so far. Three of them are (1) environmental control, (2) product liability, and (3) safety. Attention to all these is improved by an open, problem-solving atmosphere and all the other concepts advocated. Establish special goals where appropriate and recognize that attaining them is embedded in the work of attaining the fundamental goals of value-added manufacturing.

A popular view is that attention to environmental control must add cost. Sometimes it unavoidably does, but also study the matter from the viewpoint of eliminating waste. Dow Chemical pays for environmental control by making by-products of recovered pollutants.

Failsafe the release of hazardous material. Couple preventive maintenance for safety with failsafe of process quality. Not having more hazardous material than absolutely necessary is one way to minimize the risk—and the expense of controlling the material. Use it before it gets old and leaks from a container.

Careful study of customer use and simplification of both designs and processes help to prevent oversights leading to product liability problems. Attention to customer service quality helps prevent irritation that can build up to a lawsuit.

Accidents by their very nature are unexpected events. Their possibility is a part of detailed study of all operations by many people. Making as many operations as possible into standard procedures helps to prevent accidents. (Many accidents are caused by nonstandard approaches to setup or maintenance—sometimes by people inexperienced with the equipment.)

A common practice is to safeguard an operator from catching a hand in a press by putting two knobs on it that must be pushed simultaneously for the press to close. Then, the operator must stand, doing nothing else, while the press operates—a waste. The method does not prevent someone other than the operator being caught in the machine. Using light beams and photocells to stop the press when *anything* breaks the beam safeguards both people and equipment.

Besides, the most likely time to be caught is while messing around in the press, making adjustments during setup or maintenance. Eliminating adjustments during setup minimizes the frequency of such actions. Studying how to incorporate safety measures into standard procedures dealing with dangerous conditions is a big plus.

This is an example of the kind of operations thinking that should be incorporated into standard practice. Improving the practices themselves works toward the objectives of safety in the environment, safety of the customer, and safety of the employee. In the end, it is the best insurance policy and the best legal advice.

Getting Started

Everyone has to start with education. Some of it comes from books, tapes, and seminars, the number of which is growing. As soon as one is aware of what to look for, better education is seeing in action a manufacturing organization somewhat like the kind desired. Field trips are important for key people. Best is actually performing in a new way or learning by doing. (Japanese companies usually *must* reform an old organization from the inside, and they prefer sending people to live at another company for a while to become accustomed to different ways of doing things and to study role models.)

The initial education should be only the beginning of a culture of perpetual self-development by the people in a company. Some of the education is formal, but much more takes place by people regularly improving themselves by practicing at improving everything else.

Komatsu, Ltd. has been recognized for quality performance many times, and it is practicing its version of manufacturing excellence. The Chairman of Komatsu credits education for this performance. Referring to their total quality methods, he wrote:

> These programs, however, are intended to impart knowledge of TQC [total quality control] to employees by means of collective education. In order that employees may become capable of solving problems in real work situations, all personnel who participate in such programs are given a theme that may lead to the improvement of their work and, over a period of six months to one year, must work toward its solution under the guidance of their supervisors. They are called upon, without exception, to give a presentation each time a theme is solved.
>
> For this purpose, presentation sessions are held for each QC circle, staff and managers, and for each branch. In autumn every year, a company-wide QC meeting is held and presentations deemed commendable are awarded by the president of Komatsu. I myself make a point of attending the meeting every year.[1]

Different company cultures will have different ways of going about the details of this perpetual development. For all, the beginning of education is the beginning of continuous improvement as a way of life. It should be a high-profile activity. Effectiveness comes more from attitude and practice than slick training materials.

Executives also have to work up the nerve, which may take preparatory time. How to persuade. How to act. How to build enthusiasm and sustain it. How to explain without appearing an idiot or making others feel like one. Beginning education and working through doubts and fears will take key people several months. It helps if they meet regularly and talk through what they visualize happening in their company.

---

[1]Kawai, Ryoichi, "Total Quality Control at Komatsu, Ltd.," paper presented at the 30th EOQC Conference, Stockholm, Sweden, 1986, p.6.

Omark Industries was one of the first American companies to make a commitment to reform as a company. Early in its preparation, it organized a two-day program to explain everything to the board of directors. Jack Warne, president at the time, wanted to be sure the company officers were on firm ground. As soon as one begins to change a company culture, the implications begin to interfere with matters that a board of directors considers *its* prerogatives. Those who represent stockholders need to understand in enough depth not to overthrow the whole effort.

Omark and its board of directors also developed a company philosophy statement. It states what the company is and how it intends to treat its constituencies: stockholders, employees, customers, management, suppliers, and supporting communities. Many companies might consider this a waste of time, but the statement has been valuable when decisions were tough, as when the company found itself with excess employees.

Significant also is that several other American companies making a valiant effort have some kind of formal company guideline stating what they are and how they intend to behave. Hewlett-Packard, for instance, has long held by the H-P Way.

Understanding what to do seems to come easiest from understanding first how to revise the manufacturing plants. Then revise corporate activities to do a much better job preparing for production and interfacing with the plants, but keep in mind that all reform must serve the need of the customer. A rather general diagram of the approach at a factory level is shown in Figure 10–1, but each activity shown is substantially connected to those not taking place in the factory at all. (Recall the NOK, Inc., quality system in Figure 3–2.) No simple diagram can adequately cover the unification of ideas, but perhaps it will start the thinking.

From the nature of the process change come the changes in systems: production planning, accounting, order entry, and others. Some changes may be enormous while others are small adjustments, but in going through these, it is very easy to forget that the objective is to eliminate waste and simplify. Tear a process down until it is simple again and then rebuild it.

The difficult part is managing people for improvement. Reinforcement of improvement expectations is necessary. Remove inappropriate performance measures early. Add measures that promote improvement as people are ready for them. The timing

FIGURE 10–1 Continuous Improvement for Manufacturing Excellence

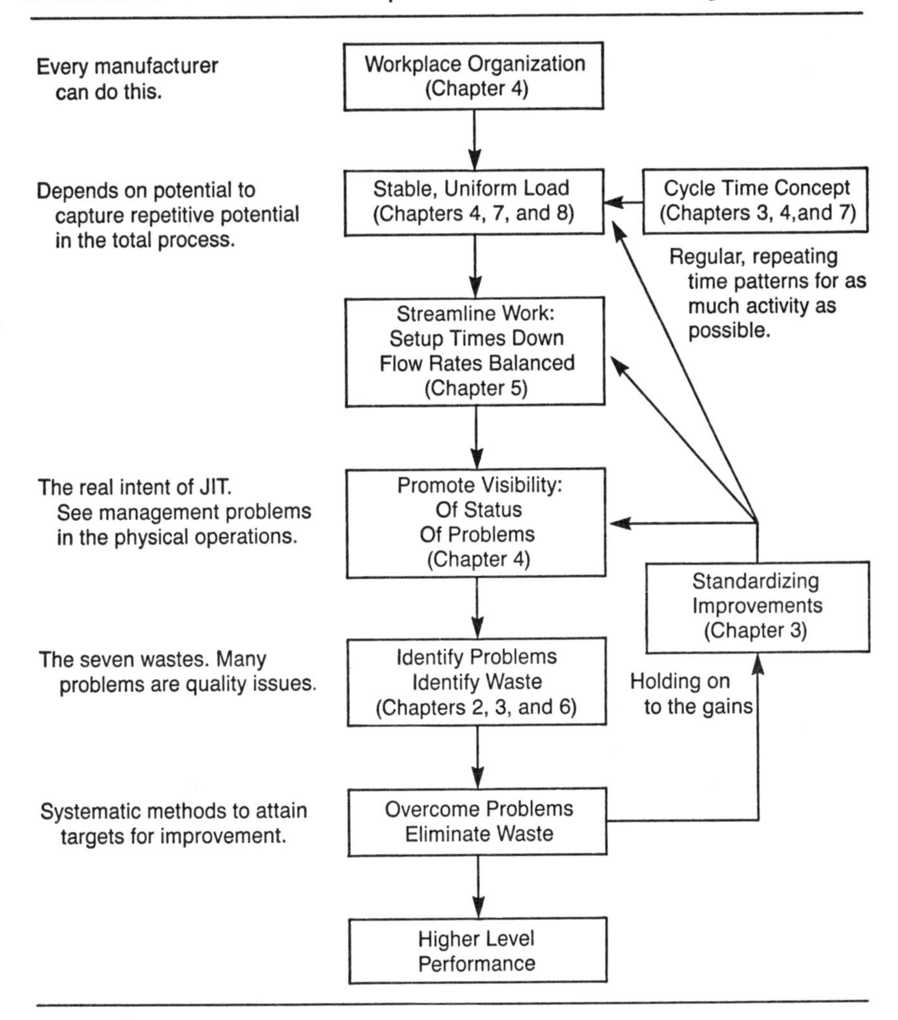

depends on the readiness of a work force to step up to more responsibility. The effort is best managed as a large program of self-development.

At the plant level, a great deal of learning is the see-and-do variety, and much improvement comes that way also. The best managers walk the floor, keep in touch with progress, and sense when people are ready for more or need a little relief. The best training is see-and-do. The best targets for improvement are those represented by something easily seen. Tell people that by next month you would like to see a layout changed; that in three months, a rework loop should be gone, with each operator correcting such errors as they occur. Follow up to be sure that progress reported is real. Someone managing a plant from a stack of paper cannot do this.

A leader at the plant level should be a good floor general, familiar with operations and people. Higher-level champions should be good generals of staff and management activity. Otherwise, the changes in engineering, accounting, marketing, and elsewhere are never made.

One of the larger beginning hurdles is credibility of the new direction. Managers and staff are all busy, or think they are, and so how are they to impose this new load of work on top of the old? Without increasing people, find a way to free the time to pay attention to the necessary changes. Look for waste in the bureaucracy and stop it. Identify some reports and activity that appear unlikely to make much contribution under the new way of thinking and discontinue them. That probably means discontinuing a few staff projects that someone considered important. Some of these only amount to better ways to sort waste. That kind of review should come early.

Organization

Each company and each plant organizes for change according to the problems and personalities therein; so no stock formula for doing it exists. Many American companies have had a rather limited view of JIT/TQ, and thus their implementations were somewhat undergunned. A company changing culture has to pursue it vigorously.

Almost all organizations have some form of coordinating or steering committee. This usually consists of an upper-level executive and the heads of major staff groups plus one or more project leaders. Their function varies from being a true guiding force to wanting to be kept informed periodically. The push for reform generally does not come from these committees but rather from the various champions in different parts of the organization.

The most successful reforms are line management driven, with the top of the line near CEO level. Most CEOs are too preoccupied with many other concerns to provide detailed guidance for internal reform. It usually falls to someone else in the top echelon to be the top management champion. An effective champion has to visit plants and other organizational groups regularly to check how change is progressing. The CEO can also do this on occasion, and should, but is usually also pressed with external commitments. At Harley-Davidson, the vice president of manufacturing took on this role. At Omark, several upper-echelon executives played the role, and so did the president on occasion. However orchestrated, the lower echelons need to know that the movement is serious and that it has strategic direction.

At the plant level, the plant manager must be a champion. Most plants at times also require project managers of one or more kinds. Some have split the effort between a quality coordinator and a JIT coordinator, which creates danger of improvements not all going in the same direction unless those two work closely together and the connections are seen by many people.

The best plant-level approaches are also line management driven; that is, the plant manager and production managers actively push the effort. Project teams report to them. Some have had very effective implementation teams for a crucial phase of plant revision.

The typical Japanese approach is line driven, but their concept of management is different. An implementation team consisting of people with considerable exposure elsewhere does a great deal of work on the shop floor itself. The team holds meetings in plain view and prods manufacturing engineers, toolmakers, maintenance, and others to modify equipment to try new ideas. Their most successful changes have come from six-month to one-year programs of crash change. Rip the plant and its systems into shreds and start over. Gutsy, but it worked. The logic is that once momentum for a change is rolling, push hard until a major milestone is reached. Some milestones: under 100 parts-per-million defects on all component processes (or under 10 ppm) or all WIP removed from storerooms.

The purpose of a period of intensity is to imbue everyone working at a plant with the thinking by working day and night until some significant breakthrough is achieved. Destroy the old way of thinking and instill a new. This requires a lot of hands-on leadership by a core group of pragmatic ideologists.

Many American companies have quality problems deeply seated in design or materials problems. They are in no position to crash a program into straight-flow production (if technically feasible) without having a lot of work-in-process inventory left, even if it is all out of stockrooms. These companies need a period of intensity which primarily attacks quality issues.

Who to put in key project management roles? That depends on the situation and the primary goal of the project. At the plant level, if the goal is quality, one needs people capable of coaching quality hands on. Needed for support are quality engineers (calibrations and measurements), manufacturing engineers (tooling and equipment changes), and maintenance (many problems rooted here).

If the project objective is conversion to straight-flow production with all WIP removed from stockrooms, two types of experience are valuable: manufacturing engineering and production control. Manufacturing engineering is important because many changes involve modifications of layout, tooling, or equipment. Production control is important to keep production coming during conversion and to devise bandages for old systems and simple beginnings of new ones. (If the cost system will be mortally wounded, accounting is vital, too.) Of course, anyone in these roles must be skilled in working with many different kinds of people. In short, pick people "who cannot be spared" from their current positions, those with enough experience, vigor, and enthusiasm to inspire confidence working with people hands on.

Depending on the situation, a project may require special service from purchasing, maintenance, material handling, or any other function. None are unimportant, and a complete revolution in manufacturing will leave none of them untouched. By the time a company or plant is changed to a new way of life, everyone will have contributed to the change.

Repeat this push for change in other parts of the organization: engineering, purchasing, and elsewhere. As the company progresses, in none of them will life ever be the same.

Hands-on is the key to a great deal of training. The best results come from dirty-hands, shirt-sleeve leadership. The best training is seeing and doing. Developing a strategic purpose may be a conference room task, but plant-level reform is a plan-as-you-work activity. Theory without action is a by-product of the cattle industry.

Line management is especially important. They put the stamp on the company. They are key to development of the line work force. They have to keep everything simple enough so that it all goes together in a way that they can handle. They must practice an old-fashioned word: *leadership*.

This translates into a management style tough in stressing performance but close to the scene of action and adept in making willing contributors. Line managers and supervisors must be developed for it. This task is not impossible, but a company does have to think about its policies for appointing supervisors. As first-line leaders in manufacturing excellence, one does not want a supervisory force largely made up of anyone who can be persuaded to move up plus a contingent of college recruits gaining experience. A company needs some of its best people in the higher-level line production positions also. Putting more responsibility *at* low levels puts more responsibility *on* them.

Line management at all levels is key to making this revolution work. A company dominated by staff tries to cover lack of authority (or competence) in action positions with expertise from the sideline. However, the revolution has trouble shifting out of the strategizing and system-building sessions. Without strong line managers, experts have trouble fitting different staff visions together into one pattern of work. The successful companies go to the scenes of action and start changing things. When line management foments a revolution, it is real—and more likely to be integrated into one piece.

MAKING THE REFORM LAST

The concepts of *total quality, just-in-time manufacturing,* and *people involvement* make up the substance of manufacturing excellence. By their implications they represent a different way of operating a company and are not techniques to be grafted onto a present existence.

A few observations about the leaders in manufacturing progress may be helpful. First, they seldom work only 40 hours a week. Some work hard and play hard, but all have dedication, and none think they face anything less than a supreme challenge.

Second, a little experience with attempting low inventories and zero defects removes a great deal of braggadocio. The manager learns to compare performance with what could be and to not be self-congratulatory on prior achievements for long. In this they are like runners who compete more against the clock than against other runners.

Third, they understand manufacturing in a broad sense: technical, behavioral, marketing, and financial. Almost automatically they begin to think of manufacturing as very challenging, long-run competition—challenging as almost any human activity short of global politics or war.

The importance of considering these things is in how to build the right kind of enthusiastic fire for manufacturing excellence. Sustained enthusiasm is necessary, and it differs from adolescent cheerleading. A large group of people must combine their talents in a fashion superior to a loose federation of independent agents. Manufacturing is often seen narrowly as a dull trade; but, done with excellence, it demands almost every talent humans possess.

MORALIZING

Preaching seems an inescapable part of persuasion on this subject, but homilies are delivered much easier than performance. Results seem to come from those too busy to talk about it. Once into continuous improvement, their education is first from doing and second from case-studying others for improvement ideas. Furthermore, those always seeking improvement have no time for self-satisfaction about it. One interesting aspect of manufacturing excellence is that if such a thing is attainable at all, it will be by those who realize that no such condition exists.

INDEX

Q-R

S

T

U

V

W-Z